The Chemistry of the Theatre

Also by Jerzy Limon

GENTLEMEN OF A COMPANY: English Players in Central and Eastern Europe *c.*1590–*c.*1660

DANGEROUS MATTER: English Drama and Politics in 1623/24

GDAŃSKI TEATR "ELŻBIETAŃSKI" (THE GDANSK ELIZABETHAN THEATRE)

THE MASQUE OF STUART CULTURE

MIĘDZY NIEBEM A SCENĄ (BETWEEN THE HEAVENS AND THE STAGE)

TRZY TEATRY (THREE THEATRES)

PIĄTY WYMIAR TEATRU (THEATRE'S FIFTH DIMENSION)

OBROTY PRZESTRZENI (MOVING SPACES)

SHAKESPEARE AND HIS CONTEMPORARIES: Eastern and Central European Studies (*co-editor with Jay Halio*)

HAMLET EAST AND WEST (*co-editor with Marta Gibińska*)

THEATRICAL BLENDS (*co-editor with Agnieszka Żukowska*)

The Chemistry of the Theatre
Performativity of Time

Jerzy Limon
Professor of English and Theatre Studies, University of Gdańsk, Poland

First published 2010 by
PALGRAVE MACMILLAN

Palgrave Macmillan in the UK is an imprint of Macmillan Publishers Limited, registered in England, company number 785998, of Houndmills, Basingstoke, Hampshire RG21 6XS.

Palgrave Macmillan in the US is a division of St Martin's Press LLC, 175 Fifth Avenue, New York, NY 10010.

Palgrave Macmillan is the global academic imprint of the above companies and has companies and representatives throughout the world.

Palgrave® and Macmillan® are registered trademarks in the United States, the United Kingdom, Europe and other countries.

ISBN 978–0–230–24111–4   hardback

This book is printed on paper suitable for recycling and made from fully managed and sustained forest sources. Logging, pulping and manufacturing processes are expected to conform to the environmental regulations of the country of origin.

A catalogue record for this book is available from the British Library.

A catalog record for this book is available from the Library of Congress.

10 9 8 7 6 5 4 3 2 1
19 18 17 16 15 14 13 12 11 10

Printed and bound in Great Britain by
CPI Antony Rowe, Chippenham and Eastbourne

Contents

List of Figures

Preface

"Theatre, above all, is the art of time ... Everything that
occurs in the theatre, every utterance, is a tool to harness
time. Time is the bed of a river, otherwise called the theatrical form."

(Romeo Castellucci, 2008)

This book has arisen out of the author's firm belief that the study of theatre,
and of artistic texts in general, may rid itself of subjectivity and be scientific –
in the sense that the opinions expressed and the conclusions drawn are
verifiable both through the internal logic of the argument, and, even more
importantly, by the performances themselves. We live in a transitional period
when the seemingly stable beliefs and attitudes of the humanities have been
undermined at a fundamental level. Academic institutions have become ideological battle grounds, and what used to be "liminal" or extreme has in some
cases become the mainstream. It seems, however, that we have also reached
the point of surfeit, and there is a growing nostalgia, also in theatre and art, for
coherence, verifiable criteria and rules. In an ironic way, the situation is reminiscent of that created in Sławomir Mrożek's acclaimed play *Tango*.

In order to discuss some of the theoretical issues connected with theatre,
rudimentary definitions and notions have to be discussed at some length.
This procedure may seem to go against the grain of today's practice, by
which definitions are denied as impossible to reach, boundaries between art
and life are blurred, and key notions (e.g. theatre, theatricality, acting) are
often used metaphorically to denote social occasions or art forms other than
theatre. The first two parts of this book cover the basic theoretical issues
connected with the art of the theatre and its perception. I have tackled the
questions I consider the most important, such as time and space, but the
book also includes separate chapters on acting and stage speech, which are
rarely dealt with from a theoretical standpoint, especially when time is the
primary focus. Moreover, in the post-aesthetic and "post-whatever" age,
theory itself has become suspect, so I have supported my argument with
sundry examples from theatre productions, often classified as postmodern,
by leading directors and companies from different parts of Europe and the
USA. In Part III an attempt is made to show how theatrical time can help in
scrutinizing and understanding the conventions applied by playwrights and
those implemented by the directors, which are not always congruous. Since
in my opinion time is indeed the "magical binder" of all things theatrical,

I have concentrated on temporal issues throughout the book, which in many ways may be treated as an attempt to grasp the complexity of time in theatre. For me, theatre is embodied time.

For many people, theatre is the form in which a literary text, i.e. drama, comes into existence, even "true" existence. Indeed, for many centuries theatre was seen primarily as "literature", that is as a dramatic text, which was brought to life on the stage. To this day many literary critics, particularly literary and drama scholars, see the theatrical performance in this way: as a scenic materialization of a dramatic text.[1] For me theatre is not drama, nor is drama theatre, although the dramatic text constitutes an important part of most theatrical productions (the basic distinction goes back to Aristotle). Many of the examples I provide in this book are drawn from the works of William Shakespeare, or relate to the productions of his plays. The choice of the Bard as the major source and evidence for a theoretical discourse is not accidental, for Shakespeare was not only a playwright but also an actor and he knew theatre inside out. That deep understanding of the theatre may be seen everywhere in his plays; not only in passages where he talks openly about the stage, as in Hamlet's well-known instructions to the actors, but also in the conventions he employs, which show an astonishing awareness of the multifarious ways in which theatre operates. My goal, however, is not to elucidate Shakespearean plays, but to use them as examples in a more theoretical approach, the conclusions of which may be applicable to theatre as such, without restriction to any writer or period. So, in spite of the fact that my examples are mostly drawn from dramatic texts created in the sixteenth and early seventeenth centuries, and their present-day productions, the book is about theatre (and not drama) in general, seen as a unique system of human communication and experience, an artistic medium. It is not about any specific playwright or period in theatre history.

Acknowledgements

The author wishes to thank Tadeusz Wolański and Jean Ward, who translated some portions of the text from the Polish. These have been incorporated into the book, often in an altered form. The remaining parts were originally written in English by the author. These were at some point sub-edited by Jean Ward and Kathy Cioffi, to whom I owe my deep gratitude. Owing to the various stages through which this book passed it is now impossible to determine who translated what, and this is why the author can only thank all of the wonderful people who at different times helped him to give the text its final shape.

Portions of this book are based on previously published materials, either in Polish or in English, and the author wishes to thank the original publishers for their kind permission to use excerpts from the following publications: "Thomas Kyd's *The Spanish Tragedy*: A Play About a Play", *Studia nad literaturami europejskimi. Księga poświęcona pamięci Profesora dr hab. Henryka Zbierskiego* (Poznań: Motivex, 1999), pp. 115–25; *Między niebem a sceną* (Gdańsk: słowo/obraz terytoria: 2001); "Shakespeare the Semiotician, or, Shakespeare Writes about His Own Art", in *Playing Games with Shakespeare*, ed. Olga Kubińska and Ewa Nawrocka (Gdańsk: Theatrum Gedanense Foundation, 2005), pp. 107–19; "Shakespeare's Soliloquies and Asides: A Theoretical Perspective", *Kwartalnik Neofilologiczny*, LII.4 (2005), pp. 301–23; "Space is Out of Joint: Experiencing Non-Euclidean Space in Theatre", *Theatre Research International*, 33.2 (2008), pp. 127–44; "A Candle of Darkness: Multiple Deixis in Roberto Ciulli's *King Lear*", *Journal of Dramatic Theory and Criticism* (Spring 2008), pp. 83–102; "Waltzing in Arcadia", *New Theatre Quarterly*, XXIV, part 3 (2008), pp. 222–8; "Theatre's Fifth Dimension: Time and Fictionality, *Poetica*, 41, H. 1–2 (2009), pp. 33–54.

Part I
The Chemistry of the Theatre

1
Reading the Elements and Compounds

Theatrum Chemicum

The title of this book is perhaps metaphorical, but it is not accidental. On the one hand, it alludes to the old tradition of inserting the word "chemistry" or "theatre", Latin *theatrum*, into the titles of learned books and treatises, which in fact were not necessarily dealing with either chemistry as such or theatre as art. Thus, chemistry often meant something learned but mysterious at the same time, an art that is not readily available to everyman. The word "theatre", on the other hand, meant a "survey", "display", "presentation" or "specimens".[1] For instance, from the full title of Elias Ashmole's *Theatrum Chemicum* (1652), we learn that the book contains "Poeticall Pieces of our Famous English Philosophers", who have written the "Hermetique Mysteries" in their "owne Ancient Language". Indeed, the book contains a number of works by divers authors, dealing with alchemy, which was the science intrinsically connected with hermeticism, secret knowledge, mystery, but was also seen as an art capable of revealing the truth about the world.[2] In some sense it is similar to how in popular opinion theatre is perceived today, as an art or medium that uses its own "language",[3] that is in some degree hermetic and mysterious and is therefore often difficult to understand, but is also capable of revealing truth about the world. Thus, the appropriateness of the two metaphors to the contents of this book is closer than it might seem. On the other hand, chemistry is a science today and a traditional brief definition tells us that it "deals with how substances are made up (their elements), how they combine, how they act under different conditions" (Hornby's *Advanced Learner's Dictionary of Current English*).

This is exactly how I propose to look at the theatre, not as an institution, but as an artistic medium governed by a system of multifarious rules or formulas. First, different substances are selected, or created *ad hoc*, such as, the dramatic text, the stage set, costumes, props, music, choreography, actor's interpretation of the role and the like. Then, following the instructions of the director, choreographer, composer of music, costume and set designer, etc.,[4]

3

within the time given, they combine or blend with each other in reactions within a stage space, which may be seen as a retort containing a binder (time). And then they (now compounds of a higher order) act according to different conditions depending on the concept and skills of the people involved, the time and place of the performance, external circumstances – economic, social, political, cultural. Sometimes the help of a catalyst is indispensable for the combinations or reactions to take place, and this is provided by *ostension* (someone describes or names something, points to something with a finger or turns his/her eyes towards it) and theatre conventions.[5] Basically, a convention is a somewhat irrational explanation through which the improbable becomes probable, the irrational becomes rational, the invisible visible, etc. This is a reminder that theatre is a sort of game, based on rules that have to be accepted by the participants who agree to take part in it.

However, the rules of the theatre are chemical only in the metaphorical sense. Nevertheless we recognize them as rules different from those that apply in the world we know. The world of the stage is bizarre to say the least; it seems tacky, artificial, crippled, unnatural and even "primitive". It seems full of pretence. Sometimes, and today more often than not, it does not seem to make much sense. A conspicuous lack of coherence, breaking the boundaries between art and life, distancing to literature and language, undermining the verisimilitude of plot and action – all these are traits of theatre today, which more than ever before seems to be merging with recent developments and the practice of art and performance art in particular (with a growing role of the media). In order to understand, we, the spectators, try to bring order to what we see and hear, to bring logic and justification to the evolving scenes. "To bring order" means simply to find the rules that govern the created world displayed on the stage, inhabited by people who behave and talk strangely. This is not always easy, especially in today's theatre practice. Knowing the rules is like knowing grammar, which enables us to read and comprehend other people's utterances. The problem is that there are no unified grammar books for the theatre. Theatre seems to be governed by many grammars, and yet there seem to be at least some rudimentary rules that generate theatre in all its possible variants. However, these rules are not innate, and we are not brought up through theatre acquisition, although many theatrical features are revealed in the games children play.

This is the major aim behind this book – to prompt the reader how to read a theatre spectacle in order to discover the rules that explain and justify its appearance in the shape given. Once we discover the rules, we begin to understand why somebody's gloves have to be light brown and not black, why there are mirrors everywhere in somebody's living-room, why there is a portrait of the performance artist Orlan hanging on the wall, and why the actors pretend not notice our, the spectators', presence, talk of landscape when there is none, the drinking of wine out of empty glasses or movement backwards. We accept as "normal" the fact that someone has the habit of talking to himself, speaks in verse, or dances on the table singing a song and nobody on the stage notices anything happening. Moreover, the performer does not seem to have any awareness of his/her own strange behaviour. We also come to terms with one wall and the ceiling missing in somebody's room, and we accept cheap substitutes for the real thing. This is indeed extraordinary, since a tawdry plastic ring can become a king's jewel, confetti can be snow, a stick a rapier, a piece of string a pearl necklace, a smudged daub can be an original Rembrandt, a plaster cast a marble sculpture, a reproduction an original work of art, and so on. A steering wheel can mean a whole car or other vehicle, a piece of mirror a dressing-table or a spotlight, a wheelbarrow an ancient chariot. But in order to understand all that we have to know the rules.

Just as certain frequencies of sound waves create phonemes, and these, when following the rules of a given language, lead to the formation of morphemes, words and sentences, so do the rules of other systems select the substances that are used and the ways in which these substances are modelled and combined (sound waves in music, clay or stone in sculpture, light and sound in film, gesture and body in dance or pantomime, etc.). In music, for instance, sound waves are organized according to rules different from the rules of the language, and the changing wave frequencies do not create phonemes but a melody line or musical chords, the formation and combination of which is (usually) governed by harmony (which is equivalent to the grammar governing a language). The selection and combination of notes draws our attention to the rules, which enable and justify their appearance in the order given. This means that music is basically autoreferential, which does not imply that it cannot evoke emotional or connotational responses in listeners. In fact, practically all artistic systems use signs for creating and encoding their messages, which means that, quite contrary to the "real" world, the modelled material matter we perceive with our senses is not equivalent to the denoted meaning of a given work or event. The meaning may be referential, i.e. referring to the world outside a given work, or it may autotellic, i.e. referring to the work itself and to the rules that enabled its formation in the form given. Much of today's art and performance art (which often presents itself as theatre) claims it is not an artefact, does not carry a message, but "energy" – a unique experience that occurs between the performer and the recipient.[6] However, as recent studies in neuroscience, linguistics, neuropsychology and cognitivism have shown, energy and experience are also activated by the so-called mirror neurons, simulated by language and/or other people's actions, utterances and behaviour. So, not only a performance artist, but also an actor in a theatrical performance has the ability to activate peculiar responses in the spectator, so that he/she experiences, at least partly, some of the feelings and emotions accompanying those of the stage figures (as signalled by the actors).

Writing about theatre, by its very nature, means getting to the heart of the complexities of a performance "text", or stage "utterance". A theatre performance is by no means limited to dialogue only and consists of multi-farious material components, or the *signalling matter*, composed of live and inanimate substances, movements and gestures, words, light and sound, creating all sorts of relationships with each other in innumerable variant proportions and combinations. These become the material vehicles of signs; since they are composed of sundry substances, often in striking, occasionally shocking, combinations, the spectators try to bring logic and sense to what they see and hear. Thus, we begin to realize that everything on the stage means more than just itself: it also begins to mean something that is not visible or audible, something that we can only imagine. We also notice that the hitherto meanings of words, objects or phenomena are seriously invalidated or altered. This prompts us to finding the reason behind these alterations, as if deciphering a secret code. Someone talks to us in a foreign tongue, we understand the intentions, but realize that part of the meaning is in the talking. Basically, a performance consists of a limited amount of signifiers,[7] selected from an unlimited material potential. There, the only limitation is human imagination and creativity, or mundane reasons like financial shortages or limited space. Moreover, some of these substances make their way to the stage as components of other semiotic systems or mediums: each of these systems in isolation, such as language, dance, music, painting, film, poetry, video, scenography, installations, can be recognized and described, but during the performance they intermingle and coalesce, creating compound configurations, heterogeneous conglomerations, amalgams of higher order, unique for the particular production.[8] Through this, they generate new relationships, hence meanings; they undergo transmutation, evading formal classification and description.[9] In other words, all the material components of what is seen and heard on the stage are contained as it were in communication vessels, and convey, blend or transfer their individual meaning one from another, thus creating configurations not to be found outside theatre. This is precisely what I call the chemistry of the theatre. Consequently, everything on the stage becomes blended into a new medium, demanding a new perception attitude and attention from the spectator.

In the metaphorical sense "chemistry" means also the ability of people to communicate, to interact with and to understand each other. The emotional and cognitive aspect of human communication is often seen and described as a chemical reaction (as in the phrase "there was chemistry between them"). So, the chemistry of the theatre involves not only the reactions occurring on the stage, but also includes the communicative and cognitive aspect of what goes on between the creators of the show, their scenic communiqué, and the spectators. And in the case of the latter that aspect may involve not only understanding but also experiencing. Naturally, there are also stage reactions occurring between the actors, and although these may influence the quality of acting (in either way), the chemistry between the actors is not part of the message. For the spectator it is impossible to determine whether there is or is not any real chemistry between the actors. Similarly, the actor's engagement in the role, his/her psychology and emotions connected with acting, are not part of the message. The spectator is unable to determine if the actor really feels suffering or joy, or is just "pretending" (i.e. acting). Even in the case of the Brechtian style of acting, where the actor seems to distance him-/herself from the enacted figure, the spectator is fully aware that this is not the actor's individual choice, but a style and manner of acting imposed upon the actor by the director. So, in fact, the actor plays an actor who is distancing him-/herself from the figure enacted, and in the former role may not be distanced at all. It may also be noted in passing that the actor does not really "behave strangely" on the stage: acting is not "artificial" or bizarre, because what we see and hear is simply the actor's fulfilment of the artistic tasks set before him/her by the director (choreographer). And that is serious and real, not fake. There is nothing artificial in the actor's doing his/her job well.[10] The introduction of all sorts of conventions that bring coherence into the otherwise bizarre world of the stage is not an attempt to fool us or to create illusion, but is the way in which theatre articulates itself. Conventionality and theatricality are the modes of expression and articulation of theatre, the meaning of which is not conventional or theatrical at all.

In theatre the communication between the director and the spectator is made possible through a system of theatre signs, unique for a given production, arranged to the specific rules of the art, which evolve before the spectator as a stage text, rendered by its creators to be "read" and experienced. The message is being articulated, but the "language" on the stage consists of material signifiers, i.e. human bodies, objects, light, words, music, etc., arranged in simultaneous three-dimensional "pictograms" and their sequences in accordance to rules that are not always clear and predictable. What is presented is often unlike the world outside theatre. Particularly striking are the time structures employed: the past is presented as the present, and that present is set within a time stream that does not seem to belong to our reality. The creation of the stage signs is usually called *semiotization*,[11] and the space where the denoted fictional world takes place will be called the *scenic space*, as opposed to the *space of the stage*, which will refer to the physical space of the material stage or the metaphorical retort.[12] These signs are much more complex in theatre than are, say, phonemes in language. The complexity derives from the fact that on the stage all sorts of substances are blended together, creating inseparable compounds, not to be found elsewhere, and the unique blends cannot recur in any other production.

Thus the stage becomes a chemical lab, in which all sorts of reactions are tested and implemented. The relation between the material substance and the meaning of a sign in theatre is neither accidental nor arbitrary – it is the result of the activities of the creators of the performance, their individual articulation and encoding (thanks to which we can also recognize the style of particular directors or of a given trend or epoch). Their activities depend therefore not so much on the selection and combination of signs already available, taken from a common stock, but on their creation from various materials, which is reminiscent of the process of the creation of neologisms in language. Here the choice does not, however, affect only lexical units or morphemes, but the kind and proportion of substances selected from various sources from which a unique theatrical signifier is formed. All sorts of substances are set within the space of the stage and begin to react with one another. As a result of these reactions, they reveal their new qualities and the ability to convey meaning, hitherto not associated with participating components. This fully complies with the definition of chemistry given above. I am speaking of compound signs, or clusters of signifiers, because in theatre none of the material elements appears in isolation: the word cannot be isolated from the actor's face, nor from his/her costume and wig, and cannot be isolated from the set, music, lights and other words that are uttered by other actors. It is these elements that create the network of relationships, which imparts to all verbal utterances (and to all participating components) a meaning specific for a given performance. The words become an element of three-dimensional signifiers which are heterogeneous, composed of sundry materials, such as human bodies, objects, sounds, light, and semiotic systems such as music, literature, dance, scenography or acting. The selection and combination is always unique for the particular production and cannot be "translated" into any other system, or found outside theatre. In other words, a performance is a sequence of compound signs, heterogeneous amalgams – if you like, created by all those responsible for the production. These blended amalgams create meaning through a network of relations, such as the rule of equivalence, based on similarity or contrast, and the rule of contiguity, or, a "theatrical syntax". This will be explained in greater detail below.

Thus, the material stage may be treated as a chemical retort, in which the director, being the experimenter, blends all sorts of heterogeneous materials from an unlimited stock, always carefully selected and combined, different in particular productions, leading to reactions, conducted under various formulas, which become visible and audible on the stage. The results of these reactions and their meaning, however, are not given in any material way and rely on the ability of spectators to "read" the chain of chemical "acts" occurring on stage and to grasp the rules and codes,[13] i.e. the formula employed, and on that basis imagine the outcome. Consequently, the spectators can only deduce the meaning of the sequences of carefully modelled scenes. But, as I have already indicated, the result of these reactions is not presented; it is to be imagined, evoked by the signals emanating from the stage.[14] It is only the fictional figures (i.e. the characters) that see the final results, but, contrary to us, do not see the components participating in the stage reactions. So, one complements the other. We, the spectators, see the reactions, the figures see their outcome; and the rule does not work the other way around.

In the phenomenological sense the performance is a sequence of these reactions, involving different components, constituting compound signifiers and their clusters, and changing with varying dynamics, pace and rhythm. These signifiers have the ability to convey information and to denote a world that is non-material, hence fictional. The spectator's task, then, is to recognize the appearance of stage reactions, involving particular components, and on the basis of their analysis, find the justification or the formula that lies behind a given reaction; this, in turn, enables the spectator to comprehend and imagine the outcome, which can then be blended with the signalling matter in the process of cognition. Experiencing the actions and emotions of the fictional figures seems to be part of that process.

The outcome of reactions is described to us by the figures who inhabit the fictional world, but this is done through the agency of the actors, through their verbal utterances, gaze, gestures, behaviour, etc. On that basis we can clarify everything that appears and happens on the stage that leads to the generating of that particular outcome. The selection and combination of the components employed in stage reactions becomes clear and self-explanatory. Thus, the "story" told in different productions might be similar if not the same, say, of King Lear and his three daughters, but it is told with the aid of different human bodies, substances and formulas of modelling. The words uttered might be the same, at least in the phonetic and syntactic sense, but their meaning is particular to a given production. Also, the ability to activate the spectators' experience of the performance varies depending on individual components of the show. Theatre is site and time specific. Moreover, it is substance specific. Theatre, indeed, is the art of blending human bodies with light, words and things, as this book attempts to show.

In other words, owing to the uncommon formulas and reactions, distinct for the medium of theatre, and particular to a given production, everything we see and hear on the stage, all the components involved, begin to react with one another, seem to blend, and consequently gain an additional meaning of something they are not. Meaning is not an intrinsic and invariable element of any text, for it depends also on the recipient, his or her individual features, intelligence, presuppositions, knowledge of the language and cultural codes used in the text, and also depends on the political, economic, social or cultural context. Since the latter changes through time, and varies from one part of the world to another, so the meaning changes. An actor's costume may be totally inadequate to what we know about its historical shape, it may also be a faithful example of fashion in a given period and place, may denote a ruler or a soldier from the historical past and a dramatic work (Richard III, for example), but certain of its attributes may refer to our reality, and create a topical allusion (meaning, say, Lukashenko in today's world), it may gain metaphoric or allegorical significance ("Christ's soldier" or an Olympian deity, for instance); it may also become an important ideological message (e.g. costumes echoing certain fashions or styles, such as camp). Consequently, the costume is never the "real thing" on the stage, even if the real thing is used (e.g. Napoleon's genuine uniform); it is a costume, and not a piece of garment, and on the stage it becomes a signifier denoting a fictional figure's garment. Naturally, the fact that the original uniform is used creates additional meanings unique for that production, but does not change the rule.

Moreover, the same material component can mean several things simultaneously or may acquire different meanings in the course of a performance's development. We see a simple brick, but we are led to understand that for the fictional figures it means a plate of delicious food, as in the famous Peter Brook production of *Ubu Roi* (1977); at the same time, the bricks mean several different things: plates, food, and bricks as such, and also they become the metonymy of the theatre building itself; at one point the actors hurl themselves against a brick wall at the back of the stage, where they hold a pose for a moment, evoking the image of an execution.[15] Thus, the actors eat bricks, being the sign of food. In French, "eating bricks" (*bouffer des briques*) means to have nothing to eat. In other words they eat nothing or fiction, both in the literal and metaphoric sense. But fiction is the actor's nourishment, so eating the bricks implies also devouring the essence of the profession, eating the theatre as art, which is the source of fiction; the latter is deepened by the fact that in the derelict *Bouffes du Nord* bricks were lying everywhere, so they were conspicuous metonyms of the theatre building. So, in some sense the actors were also eating their own theatre. Similarly, in Jerzy Grotowski's *Apocalypsis cum figuris* simple bread gains many-layered meanings (of host, human body, God's body, hunger, object of sexual desire, etc.), not to be found outside theatre. In Berthold Brecht's *Galileo* the new garments of the newly elected pope become metaphoric signs of the inner transformation of a man: the more dressed in the new robes Barberini is, the more ready he becomes to agree to the use of torture. Thus objects, words, music, human bodies and other material components of the performance reveal, through their reactions on the stage, an amazing ability to obtain and accumulate and alter meaning in ways that are not encountered outside theatre. Moreover, they reveal their ability to transfer these meanings from one to another, as if in communication vessels. This leads to semantic richness that enables the performance, which is always based on metonymy, to convey a surprising amount of information with the help of relatively simple and seemingly limited means.

But the brick from our example, even when in stage reactions it ceases to mean just a brick, does not disappear or evaporate from the stage; objectively, it is still there, and yet, seeing it, we understand that in the fictional realm it does not exist at all, and what exists is a plate of food, something we cannot see and can only imagine. The meanings attributed to objects, human bodies and other phenomena on the stage by the stage figures, as signalled by the actors, are in sharp contrast to what we, the spectators, see and hear on the stage. This causes a contradiction, for the firm relationship between a word or an object and its meaning is undermined, something that only the understanding of the rules of the theatre helps to neutralize or justify. The reactions seen and heard within the space of the stage find their results in the mental space of the spectator's mind, which, in turn, may be seen as the cognitive retort. It may therefore be said that the chemistry of the theatre relies on the establishing of a mutual relationship between the two, or perhaps even three spaces, the material or phenomenal, being the stage, the fictional, which is only implied, and the mental, which is individual and subjective, and cannot be perceived by others. Otherwise, without the appearance of these communication vessels, what we see and hear on the stage appears as ridiculous, improbable, illogical and bizarre, to say the least. Thus, metaphorically speaking, the retort of the stage requires a complementary mental retort of the spectator's mind for the chemistry between the two sides involved to take place. This is why cognitivists talk of cognition in terms of blended spaces.[16] Moreover, recent discoveries in cognitive psychology, neuroscience and linguistics have shown us that there is much more in the process of perception and audience engagement than just conscious visual and aural comprehension of the performance text. Apparently, a lot of the spectator's "embodied interaction" is unconscious. As the seminal study of Pierre Jacob and Marc Jeannerod has shown, the ways humans perceive and think about the inanimate world (generating "visual perceptions") are different from the mode humans employ watching others act (generating "visuomotor representations").[17] What this means in practice has significant consequences, for we may suspect that while watching a play, spectators react differently to the objects and scenery, and differently to the live human bodies engaged in all sorts of activities on the stage. Moreover, the visuomotor representations stimulate the spectators' ability to embody others' emotional states, even if the situation is a fictional one, and that, consequently, changes the way people feel and think.[18] What is more, scientists have discovered "mirror neurons" and their role in the ability of humans to access and experience the emotions and intentions of others by watching them act. It is therefore plausible that spectators unconsciously mirror the actions of the actors and the figures that they impersonate. So, the chemistry between the two sides is far more complex than it might seem, and has serious consequences to the ways performance perception has been so far conceived. As Bruce McConachie put it: "If people in the audience are taking pleasure from mirroring what other people on the stage are doing, the metaphorical relationship of the stage action to other actions in the world no longer seems to be at the centre of their concern. The mimetic aspects of performances may be occasionally interesting, but will not likely be the primary focus of spectatorial attention."[19] Further interpretative problems arise when the "aspects of performances" are not mimetic at all, as a number of examples provided in this book show. In those cases, it seems that the mode of perception changes from the predominantly unconscious and automatic one, to the primarily conscious and analytical. And it is clear that any particular blend might vary from individual to individual, and, consequently, we are dealing with a number of variables, not a single final blend common to all recipients. It is, however, important that the creators of a given production foresee and attempt to generate the possible blends necessary for an appropriate perception of their work within a certain spectrum.

Thus, to return to the previous line of thought, in spite of the fact that in theatre we do not actually see the final product, so to say, we perceive signals sent by the participating components, by which we are expected to deduce and re/construct the final product with the help of our mind's eyes. For this reason, theatre requires the actual presence of at least one spectator. It is due to him or her that the material space, being a chemical retort filled with light, sounds, words, objects and bodies, transforms itself into a semiotic one and consequently into a semantic and an aesthetic one (sign into significance). Here we can see that the communication-vessels metaphor may be extended to include the three retorts: the stage, fiction and the mind of the spectator, with every one of these spaces dependent on one another. There can be no theatre without fiction and without the spectator. In contrast a football match can take place without spectators, without any detriment to the essence and the rules of the game, and children can play without anyone watching. A pianist can play for himself, an actor cannot.

However, quite contrary to real chemistry, what we see and hear on the stage is the chemical process *in statu nascendi* rather than its final result, at least in the material sense. We see the substances and elements taking part in reactions, blending with one another, entering into signalled relationships (spatial contiguity or cause and effect, both being basic indexical relationships), and behaving as if they had already been transformed into new substances.[20] "As if" is the key phrase here because in theatre everything, human bodies, objects, light, music, etc., ceases to be just itself, and gains the function of a sign of something it is not, as if coated by a new quality. This indeed is an "as if" world. But, since the reacting components do not resolve into new states of matter, we constantly see them in the original form; thus, we are constantly relating the implied to the phenomenal. This explains why we may speak of an active role of the spectator in theatre. And another striking feature in theatre is that different components employed in different reactions may actually yield the same referential, though not aesthetic, result.[21] This could lead us to a common interpretational fallacy, namely, that the ultimate goal of theatre is the creation (description) of a fictional realm. This could imply that the signalling matter does not play an important role here – whatever components are used, the desired result may be achieved. However, let me stress this once again: in theatre what counts is not only the immediate result of stage reactions, but its relation to the components and formulas employed in creating it. This is what allows the spectators to notice the originality and artistry of a given production.

The fifth wall

As indicated above, the number of possible components and reactions within the chemistry of the theatre is actually infinite, for there is no system of classifying all the elements; there is no Mendeleev's periodic table applicable to the stage. It is also for this reason that a theatrical performance can be considered as a sequence of structured signifiers, preselected and unique for a given production, both visual or iconic and auditory, or more rarely tactile or olfactory,[22] whose material and meaning do not appear in identical shape and relations in any other work (nor can they be foreseen in every detail by the playwright, unless he/she is also the director). This transformation of the ordinary into "chemical" is particularly striking in the case of the seeming material inadequacy of signifiers and in the case of the elements of non-semantic systems that all of the sudden gain meaning. Also, unusual configurations of human bodies, objects, words, music and light have a powerful effect on the minds of the spectators, whose inquisitive instinct prompts them to find logic and order in the seeming strangeness or absurdity. Thus, a fragment of melody may become a sign of someone approaching, or a shift in time, a dance may reflect somebody's state of mind or a combat, and a change of colour might denote a change of space or the passing of time. In order to reveal that semantic potential, music has to cease being music alone, and a given melody line becomes subject to different rules of theatre, by which it becomes a part of a compound signifier, a blend or an amalgam, a spatial and acoustic composition of sundry materials, live and inanimate. Consequently it becomes iconic or establishes new and surprising relations, and consequently becomes semantic, capable of denoting the meaning of, say, a specific space, time, or stage figure or some other phenomena, such as it does not posses outside a given production. It has the ability to strengthen or alter the meanings of words and objects, and also to add meaning to non-semantic phenomena, such as light. Several different keys of music may, for instance, mean a particular kind of weather, war, spring, the passing of time, an approach of a ghost, a change of space, sexual intercourse, etc.[23] Usually this is done through the establishing of a unique spatial relationship between the sound or light and a human body, words, an object or other phenomenon. What has to be remembered, however, is the fact that the relationships between particular components and their functions are different in the phenomenal world of the stage, as perceived by the spectators, and in the fictional world, as perceived by the fictional figures and signalled to us by the actors. This in fact is one of the rudimentary ways in which theatre generates meaning.

At this point the initial distinction between the actor and the figure is necessary.[24] The distinction, however, is not universally accepted or even recognized, and for this reason demands some explanation. The figure does not exist in its material substance: it exists only as a network of signals and their relationships provided by an actor that have an impact on the spectator's psyche and have the capacity to evoke the mental, imaginary and cognitive schemata by which, as a recent theory would have it, a conceptual blending process is initiated. In fact, as indicated above, it is also the creation of all sorts of relationships that occur between a given actor and other actors, his/her costume, wig, make-up, music, language, the stage set and the like. The spectator takes all of these factors into account, consciously or not, in the mental process of conceptual blending, in creating models of fictional worlds and the figures that inhabit it. The idea of "conceptual blending" is taken from recent developments in cognitive studies, which treat "concepts" as perceptual categorizations necessary for the way we perceive the world. These are not innate, or given, but gain neuronal structure in human minds through embodied interactions with the environment. And looking at the live actor impersonating a fictional figure provides not only specific interaction, but a complex blend of the real and fictional. As cognitivists put it, the spectator can be in or out of the resultant blend, and can also "live in the blend", often below the level of consciousness. To some extent this perceptual mode can be controlled or stimulated by the actors, who may strive to adjust their style of acting in such a way as to foreground their skills, or, say, the emotions of the fictional figures.

In other words, the figures are created by material signals that never reach them in the fictional world and are communicated directly to the spectators (omitting the senses and consciousness of the figures). It has to be stressed therefore that the actor is a live human being for whom theatre is a profession or hobby, whereas the figure is a fictional creation of the former. In other words, the actor is both the co-creator and the material substance of a sign of a figure, which is immaterial, and basically a mental construct ("such stuff/ As dreams are made on" – William Shakespeare, *The Tempest*, 4.1.156–157). We have to keep in mind, however, that the figure is often much more than an actor is capable of creating on his/her own: the fictional figure is a synthesis of the various relationships between fictional and material and verbal substances of the performance. What this means is that the figure is constructed not only by what the particular actor who creates it does and says, but also by other factors, such as other actors' utterances, behaviour, costume, make-up, light, music and the like. Part of that information contributes to the fictional being, whereas part contributes to the techniques and aesthetics of the acting and is not perceived by fictional figures. It seems therefore that it is worthwhile to make yet another distinction between the figure (of, say, Hamlet) and the stage figure (an actor playing Hamlet). The first one is a fictional being that is being created (the signified), the other is the stage signifier: it is the actor's body that has been given a new function to mean something it is not; but, as I said, the figure is a synthesis of information generated by other signifiers, too. The figure is the fictional construct; the stage figure is the actor at work.[25] Thus we may distinguish different ontological phases in the rise of a "character": a human being → an actor → stage figure → fictional figure. When we evaluate acting, we do not have Hamlet the Prince in mind, but the actor who has created the former. Hamlet the Prince does not have any attributes of a professional actor, even though he is complemented by Polonius for his delivery of a speech from a play. It is the actor who plays Hamlet who conveys all sorts of information and signals concerning his/her art, the way he/she constructs the figure. Also, many signals and considerable amount of information generated by other stage components (other actors, language, the stage set, music, light, movement, etc.) contribute to the aesthetics of acting, and are not perceived by the fictional figures (for these signals are not part of their world).

Thus, the magical chemistry of the theatre enables the actors to signal their transformation into whomever or even whatever the author has intended (they can only not remain themselves, i.e., actors cannot impersonate themselves in the here and now of the spectators):[26] in the created world of fiction the body of the actor can be a sign of a memorial or a coathanger or a chair, or it can be an animal. The fictional figure does not even need to be enacted by a human being – it can easily be created by a puppet, although these have to be animated and have to reveal at least some individuality: what is really needed is their signalled awareness of their "I" and of "here" and "now", by which they establish their *deixis*.[27] Also, there is a tendency in the so-called postmodern theatre to detach deixis from the body of the actor, as in works of Richard Foreman.[28] The fact that such transformations – often drastic in today's theatre practice – ("o'erstepping the modesty of nature" as Shakespeare would say) can irritate us or will not be understood is another matter altogether, and is not the subject of this account.[29] The only important thing is that we agree, as spectators, to accept the fact that there is someone who reads the phenomenal world of the stage in a different way, as elements of his/her true reality, which leads to the ontological split of everything on the stage and may lead to what traditionally has been described as a temporary willing suspension of disbelief. Consequently, we agree to decipher and interpret everything which is being shown us on the stage, including the actors' bodies and costumes, their verbal utterances, music, lights, properties and scenography as signifiers and sequences of signs, creating in our mental perception a model of a different reality (which is partly possible thanks to the cognitive schemata by which we perceive the world). We do not have to see that reality "with our own eyes" – what suffices is our conscious ability to accept someone else's perception as valid in the created, illusionary world (which we then try to compare with our own experiences and understanding of the real world). The fictional world created by the stage signifiers is not therefore a mirror image of anything (we cannot see fiction); at best it could be a model of a certain reality and the carrier of a certain communiqué or experience, which the creators of the performance wish to present to us, the spectators. This model can only be an approximation of extra-theatrical reality ("as if a mirror", as Shakespeare put it: please note that he says *as if*), it can never be identical to it. It is ontologically distant and it cannot be perceived by our senses. As I have already indicated, it can only be created in our mental processes as a cognitive model or structure. Thus creating fiction is a mode of articulation, and not the ultimate meaning of theatre.

In theatre the material vehicles of signs very often do not look like the real thing. For example in Robert Wilson's Copenhagen production of *Woyzeck* (2000), a wooden stick became a sign of a wine glass – in this case the total lack of similarity was compensated for by the gestures of the actors (they "clinked" their sticks) and by sound (at the moment of clinking the sticks emitted sounds characteristic of glass). Since the material of the object which we see in this meaning contradicts our life experience, and also our visual memory, knowledge of physics or awareness of cultural codes, there arises the hiatus or discrepancy between that which is seen and that which it means. In this way the object becomes a conventional sign, conspicuous through its unusual material substance, of that which in reality is not seen. It could generally be said that during a theatrical performance everything on the stage – to a greater or lesser degree – creates such a hiatus, and in so doing reveals its new function, which the scenic figures, however, do not notice (with the possible exception of scenes taking the form of theatre-within-theatre). On the basis of various relations between its material components the performance constructs itself in front of our eyes and ears and reveals the rules that made the new structure possible. In other words, meaning in theatre is generated through the clash of at least two conflicting models of perceiving reality, the figures' and the spectators'. The word or gesture commands us to see that which is not on the stage. Noises offstage command us to see spaces, which cannot be seen. An actor's gaze may be the sign of spaces we cannot see or of an approaching danger. A simple sound (not only the spoken word), light, gaze or gesture have the ability to create (denote) or evoke a picture, even a spatial one. This phenomenon can be called a *scenic synaesthesia*. William Shakespeare would instruct us to look with our ears, as Lear instructs the blind Gloucester to read the (non-existent) challenge in 4.6.

The more discrepant the two spheres are, the more theatrical and conventional the stage world appears. The less effort it demands on the side of the spectator, the more similar they become (as in the mimetic stream), the more lifelike the performance is. A total merger, however, can occur only as an illusion. And by illusion I mean a situation when the spectator begins to perceive the world on the stage in the same way fictional figures perceive it (put precisely: how the fictional figures created in the individual spectator's mind seem to perceive their world). We see the visible world on stage as if through his or her eyes, which means that instead of the stage we see an interior of somebody's bedroom, a hall in a castle or an orchard, instead of actors we see "real" people who are not wearing costumes and wigs, but have beautiful garments and hair. If at any moment of a given production we see and hear Lear on the stage, it does not mean that he really exists – what it means is that we have fallen under momentary illusion. However, in order to comprehend and perceive the play correctly, we do not have to fall under any illusion; the latter is not and cannot be a criterion of the artistry of a given production (think of children for whom the stage often becomes reality). Succumbing to illusion to a great extent depends on the desire (one may assume this is usually unconscious) to mitigate the cognitive dissonance through a periodical assumption by the recipients of the stage figures' point of view and their modes of perception of reality. Obviously that is possible only to a certain degree, because we rarely fully forget that we are in the theatre, although, it could be supposed, this is not impossible (especially in the case of naive spectators).

At any rate, illusion should be seen not as an artistic goal to be achieved in theatre, but, rather, a psychological state showing misapprehension of theatre.[30] Of course, the illusion that theatre evolves by itself is a convention. We are constantly conscious of the fact that someone has created all this and – with the help of many people, objects and technical apparatuses – is showing us various bodies, gestures, scenic utterances and actions constituting a communiqué or event in the display case or chemical retort known as the stage. The very lifting of the curtain implies that someone invisible is lifting it – so besides what is visible on the stage, there are some other forces, which make everything move. This means that we begin to suspect the existence of some mechanism, which is the driving force behind the performance, and that everything we see is being shown to us by someone whom we cannot see. Even if there is no curtain, as in the Elizabethan stage, we see the actors enter and exit through the same pair of doors, each time signalling that they are somewhere else, in a different space and often at a different time. We also recognize the fact that quite often the entrances do not necessarily mean that in the fictional world someone is entering: the scenic doors do not necessarily have to denote doors in the fictional world, and the actor may indicate that he or she has been waiting for someone in that very place for two hours. Thus we can easily guess that the behaviour of the performers and the texts uttered by them are the realization of a particular scenario, created by someone else. Just as someone composed the stage space, so were the costumes, wigs and properties. Even though some of the latter might look like the real thing, the whole setting is usually highly "artificial" when compared to the phenomenology of the world around us. Please note that it is artificial only when that comparison is made. Otherwise, the seeming artificiality is absolutely natural for the ways in which theatre articulates its art. Quite contrary to film, theatre does not want to appear as an illusion of reality. It constantly draws the attention of the spectator to the fact that what is seen and heard on the stage does not carry the same meaning in the created world. Let me repeat once again: the semantic clash of the two spheres, the phenomenal and the fictional, constitutes the basic way in which meaning is generated in theatre. In order to achieve that mode of expression, theatre as art requires the articulation of at least two different modes of perceiving reality: one that perceives and describes the phenomenal, and the other that perceives and describes the fictional without, however, recognizing the fictionality. The first of these modes are the senses and the perceiving mind of the spectator, the other is provided by the actors, who, through their gaze, gestures, behaviour, etc., describe the world as perceived by the fictional figures beyond what I call the fifth wall.

So far, in scholarship only the invisible fourth wall has been distinguished, separating the box stage from the auditorium, often identified with the bourgeois theatre and its aesthetics. The abolition of this wall is often seen as a sign of breaking through fossilized conventions. However, it seems reasonable and necessary to introduce the concept of the fifth wall, which in theatres of all kinds separates the material substance from the fictional sphere, which separates human bodies, props, costume, music and the like, from what all this matter denotes in the fictional realm.[31] In addition, the fifth wall marks the division between the two spheres governed by different laws of physics, and that includes geometry (space) and – most importantly – time. The spectators do not experience the fictional present time in the way they experience their real present: the former may only be described to them by signals from those who allegedly experience it, i.e. the fictional figures.[32] The fact of describing time verbally is not unusual, because time is always lexicalized in terms of space and motion through three-dimensional space. Thus, the experience of real time might be contrasted with the fictional time, being an intellectual construct (there are no stage properties or emblems representing fictional "time"; there are no clocks that can measure it);[33] but in theatre the spectators might experience the phenomenon of temporal illusion, by which they share the experience of time with the figures.[34]

So, during the performance, the world of the stage reveals its dual nature, with the concomitant division of time and space (and dual ontology, so to say), followed by the conspicuous emergence and juxtaposition of two models of perceiving reality, that of figures and that of spectators.[35] Consequently, time in the performance becomes split: it is the real time of the performer, who is a live human being, and the created fictional time of the figure (which may seem occasionally to overlap). In fact, we are dealing with not only split time, but also with two presents: one is the present time of the live performer and the audiences, the other is the present time of the fictional world, which might be labelled *agreed* or *transferred present*. This is why I am speaking of the theatre's fifth dimension, which results from the dual present time. The latter forms the *sine qua non* of the theatre, which explains why so much space in this book is devoted precisely to temporal issues.

The dual nature of everything in theatre means that in consequence we have at least three semiotic orders in constant play, each characterized by its own peculiar hierarchy of functions (iconic, indexical and symbolic, if we stick to the distinctions made by Charles S. Peirce), anchored to at least three different streams of time. The first of the aforementioned semiotic orders may be called scenic or phenomenal, the second, denoted or referential, and the third conceptual or cognitive. This creates a triad, which is characteristic of theatre.[36] The constant confrontation of the first two results in the appearance of an imaginary construct in the mind of the spectator, which is also governed by its own semiotic order, set within a different time stream that we may label psychological. This may be treated as an imaginary or cognitive order, and is, in fact, a sort of synthesis resulting from the clash of the former two with the spectator's individual features, the qualities of his/her mind, psyche, knowledge, experience, etc. Since it is a mental construct, it is subjected to the psychic rules that govern cognition, which, in turn, may be distinguished in spatial and temporal terms. Human memory, which enables the spectator to grasp the whole performance and see various relationships between heterogeneous components of the signalling matter, may be compared to a cinematic montage, which records reality in hundreds and thousands of fragments, glued together to form a sequence, often metonymic in their appearance. The performance, as memorized by the spectator, loses its essential feature of a temporal continuity, of evolving present time, and becomes a form of narrative recollection, an edited, mainly visual, version of events past. Thus, in each of these orders the model of the world will take a different shape and will be governed by different laws which also include time. Also different will be the threefold relationship between the signifier, its substance, physical shape, appearance and characteristics, and the denoted object, figure or phenomenon, and the imaginary construct in the mind of the spectator. It is important to remember that the relationship has stylistic, semantic and aesthetic traits.

The imaginary construct, created in the mind of the recipient, is always confronted with what may be seen and heard on the stage. The continuous development of events and the evolving images of various compositions and substance on the stage are confronted with the imaginary schemata in the mind of the recipient and create a sort of feedback: the one influences the other, from the beginning until the end of the performance. This is corroborated by the findings of cognitive psychology, which suggest that experience derives from "an on-going perceptual process (the present), which integrates perceptual input with, and hence modifies, schemata stored in memory (the past). The modified schemata are in turn used in order to generate expectations (the future), and hence to anticipate new perceptual experience. This represents a continuous perceptual process of updating successive perceptual information to which organism has access."[37] Until the very end, the evolving events, utterances and images have an impact on the spectator's memorial "edition"; and vice versa, the spectator's cognitive responses and memorial montage (subjective in nature) have an influence on how the ensuing performance is individually perceived, what the mental editor cuts out and what is preserved in memory. In this way an individual network of relationships among various components of the performance is established, and this may, but does not have to be, similar to or congruous with the one intended by the creators. These reactions between different semiotic orders include the language, which in its mental recording takes a different shape. We do not memorize the whole text, but only fragments of it. However, the remaining parts are not lost: they are converted into cognitive schemata (blends), which do not have to be verbal at all, by which a given performance is experienced, interpreted and remembered. It may also be observed that the congruity of the mental creation, i.e. the imaginary order, with the ensuing material components of the performance text will result in positive responses from the spectator, while lack of congruity inevitably leads to disappointment, bewilderment or even total rejection of a given production. Naturally, artistic texts tend to avoid predictability and intend to surprise or even shock the reader/spectator with solutions not to be found in life or in other works. The tolerance of the "unnatural shocks" that extreme art and theatre give their recipients is highly individual, although it may be influenced by vogue, stardom of the creators, or social, political or media pressure.

2
The Chemical Formulas
of Blending

Theatrical markers

It may be said that theatre is an artistic medium, which partly presents and defines itself, explains its own rules, and – being at least to some extent self-explanatory – justifies the existence, selection and modelling of all its constituent substances and components. In order to convey its message, theatre has to find or create a space where it can present and express itself as a medium. It does not mean that a theatre building is needed for that purpose. For this is not simply a matter of the architectural space, because a theatrical performance does not have to take place in a building which we recognize as a theatre-institution; to all intents and purposes it could take place anywhere. Furthermore the background to the performance does not have to be the stage set – it could be a real street, and the actors do not have to wear historical costumes – they could wear clothes which are no different from those of the passers-by. So how do we know that this is theatre and not, for example, a political demonstration or an advertising stunt organized by a local department store? Sometimes this distinction is not easy, as, in contrast to other works of art, theatre can become similar to non-theatrical reality because usually its substance in a visible and recognizable way belongs to reality and is material and three-dimensional, and sometimes may even seem to have an "extension" or continuation outside the stage and theatre spaces. We perceive this world as really existing, at least in the material sense, and during the performance we recognize at first sight its elements, such as, for example, the "real people" appearing on the stage or the various usable objects. Sometimes these are elements of nature, particularly in open-air performances. So, there must be something else that distinguishes theatre from non-theatre.

Thus a theatrical performance has to signal its distinctive features to the spectator(s), which is essential for adequate cognition. It has to demarcate its space and time, so that the spectator may be able to distinguish the performance from other phenomena in the surrounding world and may switch to a specific mode of perception. And we have reasons to believe that the perception of theatre involves cognitive responses different from those which are employed during the process of perceiving the natural world. To achieve this, *theatrical markers* are employed. These are not always obvious, especially in today's theatre practice, where we witness a tendency to blur the boundaries, to leave the spectator or witness bewildered as to what exactly he or she is watching. Differences in the levels of this awareness and in the method and ways of perceiving the world are an important factor in steering the recipients' reception of what is shown on the stage (on them are also grounded the phenomena of tension, suspense, irony or even humour), and are in fact one of the basic methods of generating meaning. If the fictional figures read and perceived the world in the same way as we the spectators do, no theatre would be possible. This explains why literal modes of presentation are detrimental to the essential nature of the art of the theatre, and it takes talent and imagination to employ them in an artistically fruitful way.[1] Behind advocates of literal modes of representation lies the naive belief that the theatrical fiction with its "artificiality" is less "real", and, being highly conventional, mirrors the social conventions rather than reveals its ability to change human beings and their awareness. The same argument is used against traditional acting, with its alleged pretence considered the major sin of modern or bourgeois theatre. Again, behind this view lies a naive belief that the lack of pretence heightens the persuasive, artistic or intellectual powers of an utterance, message or event. In my opinion, acting has the ability to convey much more information than natural behaviour.

The more the theatrical performance emphasizes its difference in relation to the world outside, the more the specific hiatus between the phenomena of "our" reality and the phenomena shown on the stage will mark its presence in the reception of the performance. It transpires that our hitherto infallible deciphering codes and the cognitive schemata by which we perceive the world do not fully suit and apply to the stage world, the result of which is the certain obscurity – or at least difference – of what we are watching. However, on the whole it may be said that in theatre everyone and everything signals to be something different from what it seems. The figures enacted on the stage, however, do not notice this seeming artificiality and pretence, are not aware of the dual present time, which constantly emphasizes one of the fundamental features of the world of fiction: the material substance of a sign (object), as we perceive it, does not play a significant part in the fictional realm, and our knowledge and cognition of the world differs, often to a great extent, from the knowledge and cognition displayed by the figures. What for us is thin air may be of visible substance to the figures (as, for instance, the "fourth wall"). And *vice versa*, what we see and hear remains invisible and inaudible to the figures (e.g. an actor who impersonates a ghost remains perfectly visible to the spectators and yet the figures do not see it). This has to arouse our sense of surprise. We recognize the fact that what is seen and heard within the boundaries of a given space, and within the time given, means also something different from its literal appearance. So, in order to understand theatre, we need to know its rules and the ways it marks its difference from the world outside.

It has to be noted, however, that the process of learning in theatre is not similar to language acquisition at all, for there is no such thing as a unified theatre grammar, lexicon or pronunciation. Only a handful of basic rules has to be observed. Every production uses different signalling matter, and is often composed according to the rules that need to be recognized as such, rather than belonging to a common stock. However, every performance contains a "prompt-book" and "instructions" on how it ought to be read and understood. In this way theatre shows how it wants to be perceived and interpreted. The important thing is also the fact that we, the spectators, do not have to respond in the same "language": our task is to transform theatre signs into units of cognition, to blend the fictional with the phenomenal. Since information contained in the performance is revealed in time until the very end, the "dialogue" between the author(s) and the recipient continues until the fall of the curtain. The aim of the former is to convince the latter that the form of the show, as modelled and presented on the stage, is not only justified by external circumstances (such as current vogue, aesthetics or politics), but also by the set of internal rules that have been implemented and by the choice of substances that have been used. Please note that the act of communication, or the process of cognition in theatre, is in most cases highly independent of the close physical relationship between the performer and the spectator. Again, it is naive to think that the abolition of the physical distance between the two will enhance the quality of the performance, heighten its intellectual or artistic scope, or stimulate cognitive powers in spectators. It may however lead to a specific kind of experience. Thus, the "speaking picture" of the stage, through which the author speaks, is constantly engaged in a dialogue with the silent brain of the spectator.

Returning to the previous line of thought, it may be said that the most immediate way in which the theatre marks its otherness is through the seeming material and functional inadequacy of everything visible and audible on the stage. By "inadequacy" I mean the breaking of reception habits and conceptual schemata by which we perceive the world around us. In theatre, the material inadequacy becomes, paradoxically, a systemic mode of constructing and modelling the performance text, so what is adequate for the director to express his/her artistic goals, may look highly inadequate when compared to the outside world. With the possible exception of live human bodies (of the actors) and some of the props, what we usually see is not the real or complete thing, it is often a cheap substitute made of inappropriate materials, or it metonymically stands for the whole; what we hear does not seem to refer to what we see, but to another world that is said to be real, not "fake". But we do not find that world on the stage. What we see and hear are words, objects, human bodies, engaged in all sorts of relations or "reactions". We see that the crown is made from paper, the swords from wood and the tree or rock from papiermâché. But at the same time we understand that to the denoted fictional figures the crown is real, as is the rock and the sword. When a door is slammed on stage, the cardboard walls shake; the tree outside the window is painted; the actor is pretending to be someone long dead. Furthermore his/her actions and gestures are exaggerated and often unlike those we use in daily life; the actor speaks his lines out towards the audience, while we know from the context that he is looking at the wall of a house, a garden or a beach – all invisible to us. Even if the set, the costumes, props and other material paraphernalia of the performance (along with the live bodies of the actors) are as if taken from somebody's house, kitchen and wardrobe, we notice all sorts of weird things and actions. People behave strangely, as if not noticing our presence, as if living in separate worlds and not even aware of our existence; they talk loudly, live in rooms with no ceiling and lacking at least one wall; time in their world seems to pass with a different pace, and everything seems to be taking place elsewhere and at a different time, often long past, and yet they pretend as if it were taking place here and now. They describe the "inadequate" world on the stage as the materially adequate world they live in.

We can immediately notice that although the language spoken by the figures is, in most cases, admittedly the articulation of a human tongue, it is a somewhat different one; it can also astonish us that the figures speak on a daily basis in verse or sing and dance in public or talk to themselves under the influence of emotional impulses.[2] Moreover, we may notice with surprise that very often, with rare exceptions,[3] within the fictional world nobody notices the verse or the dance and the song. Our astonishment can also be evoked by other things, for example by the fact that, unmoved by the presence of the spectators, the figures declare their love for each other, reveal secrets, plot or organize revolutions and *coups d'état*. They talk about the world around them, obviously created by somebody, as if it were the real world, they see things we do not see, and vice versa, do not notice things that are obvious to us (such as the curtain or the lights or the material inadequacy of the stage set and props), which somewhat turns upside down our deeply rooted beliefs in the permanent connection between language and material reality.[4] We understand that the words used on the stage do not refer literally to everything we are witnessing (and we are witnessing a theatrical performance which creates a fictional realm), with the possible exception of lines in which the actors/figures seem to reveal some awareness of the theatrical situation they have found themselves in (as in asides). In this way the language or stage-speech as used by the actors on the stage seems to be predominantly a sign of the language used by the fictional figures, and, paradoxically, it is only that implied language we cannot hear that denotes the other realm, refers to a notional world. However, the stage speech is additionally and inseparably blended with the material context, with furniture, pieces of painting, architecture, sculpture, music and dance, obviously in various combinations or hierarchies. This creates a conspicuous discrepancy between the language as used by the actors and the visible material, spatial context, which obviously belongs to our, the spectators', world. As we can see, also the language on the stage reveals its dual nature, its theatricality.

For within the chemical retort of the stage, bodies, objects and other phenomena are given a new feature of *theatricality*, which may be understood as a dual function: the phenomenological appearance on the stage along with the function and meaning usually, though not always, known to the spectator from everyday life experience, and the acquired or denoted shape, function and meaning in the fictional realm, as described and signalled to us by the figures via the actors, which are not congruous with the former.[5] Only the figures, as signalled to us by the actors, inhabit the world beyond the fifth wall and can convey us information about it. I am stressing the importance of the figures' perspective, because their perception of the world, their understanding of the language and, generally, their reaction to and awareness of the situation they find themselves in, is usually strikingly different from the awareness, understanding and perception of the spectators. Several planes of reference are at play here, and playwrights and directors are certainly aware of the fact that this provides an additional mode of signification. A word, a phrasal verb or an adage might have specific meanings in the created world, but, simultaneously, have the capacity of evoking certain associations and connotations in spectators.[6]

It can therefore be said that the theatrical performance is a flowing sequence of scenes emanating bundles of signals, amalgams of sundry substances and phenomena, some of which are created *ad hoc*, while others initially belong to other systems (e.g. music, dance, pieces of sculpture or painting used in the performance). "Initially", because these amalgams – being the result of the process of semiosis – become subordinated to a new medium, which, on a similar principle as the grammar of a language, through a series of rules defines what belongs to the performance and what does not; in other words the rules and the selected components establish new relationships between newly formed, compound signs and also – through the creation of a meaningful structure – define the boundaries, the framework, of what constitutes the world created by theatrical means, and what remains empirical reality, the surrounding space. Furthermore, as already indicated above, the preselected substance of theatrical signs is specific and unique for a given production and it cannot be repeated in any other. In contrast to the limited resources of phonemes or the lexicon of a natural language, the resources of a theatrical performance are unlimited and do not form a common stock for all the creators.

In literature there is no such thing as the inadequate substance of signs: all words are transparent (they do not create meaning by their physical or material shape, with the possible exception of calligraphy and some experimental writings) and at a given time are phonetically and graphically the same, whether used by James Joyce or Ernest Hemingway or T. S. Eliot. Their meaning is not determined by their shape and substance, whereas the latter is exactly the case of theatre. The meaning of William Shakespeare's *Hamlet* does not change with changes in quality or size of print or paper; in theatre it does change in every production, for the human bodies and materials used are different, and so is their modelling. Moreover, in theatre we may find an almost unlimited number of material substitutes of signifiers that denote the same thing (in language the number of synonyms is limited), which means that linguistic signs may be substituted by objects or other material components of the performance (e.g. non-verbal sounds or music: in a similar manner a refrain of a hit song, played out instrumentally, carries with it the lyrics). However, substitution does not mean synonymity or aesthetic sameness. Quite contrary to literature, where in most cases the actual shape and colour of fonts or letters bears no relation to the semantics of the text, and these fonts and letters usually become transparent during the process of reading, in theatre the substance of signs is of great importance: not only does it draw attention to itself, to its otherness, by which it ceases being transparent, but it creates all sorts of relationships with other substances of a given production and with the elements of the implied denoted world (and, of course, directly or indirectly with the elements of our empirical reality). This in itself is a proof that the artistic aim of a production is not the description of the fictional world only, but also to establish meaningful relationships between its particular components or signifiers.[7] This also explains why in this book the significance of the material substance in theatre is given so much attention.

A question may be raised: if the substance of the stage world is highly inadequate (even the bodies of the actors are not always adequate),[8] how do we know what it actually denotes? Sometimes we are able to decipher this ourselves, especially in mimetically oriented plays, or for instance by the functional use of an object by the actor: if he puts a paper crown on his head, we understand that this is the real thing in the fictional world, where the crown is real, and that he, although wearing jeans and a T-shirt, is the King. How do we know that the crown is real? First, owing to its iconic resemblance to the crown, and also through its contiguity to the actor's head and the manner he is wearing it. Secondly, we know that because figures inhabiting that world tell us so: they treat the crown as the real thing (strictly speaking it is the actors who impersonate those figures who signal to us how the figures perceive the crown).[9] If they do not, it could mean that they are engaged in a game of some sort, or that perhaps the action takes place in an asylum. Consequently, since we know that the crown is real and the actor is a sign of the King, his jeans and T-shirt become signs of the King's appropriate garments, since in the theatre we, the spectators, follow Shakespeare's advice and "deck our kings with our thoughts" (Prologue to *Henry V*).

In other words, the jeans and the T-shirts enter into a chemical reaction with the crown (through the agency of the actor's body, words and gestures), and create a relation of spatial contiguity with each other, which, in turn, conveys, as if in communicating vessels, the meaning of the crown onto the garments worn by the actor; the reaction may be supported by a verbal catalyst, which is ostensive in nature, and the behaviour and utterances of other actors. Moreover, even if the stage is almost bare, thanks to the paper crown and the way in which it is perceived by stage figures, we understand that it becomes a hall in a castle, and the simple stool becomes the throne. Thus, a paper crown becomes a marker and a metonymy of time and space, bringing along all sorts of secondary meanings concerning the figure, his/her role and position in the society, etc. In Henry Chettle's gory *Tragedy of Hoffman* (1602?), there is an iron "burning" crown used as an instrument of torture; at the same time the crown denotes space (a cave) and, symbolically, the protagonist's assumed duty to take revenge for the death of his father, whose skeleton, wearing the crown, is hung in the cave. The same object might have had referential function when staged in 1602. The same, seemingly empty space may also denote a battlefield and the stool may be a sign of a horse on which the actor will gallop around the stage. The rule applies also for all those instances when the material substance of the theatre sign is totally inadequate in relation to the empirical reality and does not reveal any similarity to anything we know. It is through the way that the figures "read" and verbally describe these objects or phenomena that we are able to grasp their intended, though only implied, meaning. It does not mean that we begin to see something as if in hallucination: through the continuous signals emanating from the stage, we blend the input spaces and time to convert the fictional worlds and into cognition. Fiction becomes a mental construct, a cognitive model of the world. However, fiction is not the true meaning of theatre signs – the latter, let me repeat, is achieved through the relationship between what is denoted (the real crown) to that which denotes it (paper shaped as a crown). The ultimate meaning would have been entirely different if the crown had been made of thorns, a chamber-pot, or if it were materially non-existent.

All of this means that theatre signifiers do not necessarily have to be iconic in order to mean specific objects or phenomena. They may be indexical or mixed in nature, and the vague or even non-existent similarity may be substituted by spatial contiguity (an actor holding an invisible dagger), or by the logic of the cause-and-effect chain of events (an actor bouncing off an invisible wall). An object in its material shape does not have to resemble in any way that which, thanks to scenic clarification, it is to signify on stage. Its spatial contiguity to other substances co-creating a compound theatrical sign means that a pan-pipe can signify the "mendacity" of a figure, a simple brick might be roast meat, which in addition the figures will consume with relish. And it is this imputed or ascribed similarity and contiguity, sometimes astonishing in the semantic and functional distance of the concept of meaning from the substance, which decides the exceptional feature of compound theatrical signifiers. In other words, the particular heterogeneous elements of the substance of compound signs can take over the qualities of similarity or contiguity, one from the other, as if in communication vessels, thanks to which the theatre has an exceptional ability for changing those relations, between the substance of a sign and its meaning, between the signifier and the signified. What is more, particular components of theatrical spatial signs have the ability to denote the same signified, which could lead to tautology, especially when they appear simultaneously. But it does not, because these signs are different in the material sense, and their selection and combination leads to the growth of their artistic quality and semantic power rather than to tautology. This is why we may talk of tautological strengthening of generated meaning, and – as cognitivists would have it – the theatre's enhancement of the ability to activate the spectators' state of experiencing the feelings of the figures, as simulated by the actors and other components of the performance. Moreover, one substance of a compound signifier has the ability of becoming its metonymy. The paper crown can be a metonymy of a particular king, royalty in general, of a hall in a castle, of the whole castle, of the Middle Ages and so on. Let us add that the dramatic author often signals the potential of a symbolic object or word by the title of the work (*The Wild Duck*, *Arcadia*, *The Cherry Orchard*, *The Balcony*, *Much Ado About Nothing*[10] and so on). However, we must keep in mind that the symbolism and the metonymy is often the creation of the director.

Absorption of meaning

The phenomenon of transmission of meanings from one sign to another, which we call the *absorption of meanings*, requires a further commentary. So, not only the same object or actor can change its own sign nature (to be this or that), but can also transfer meanings, as if in communication vessels, from one to another. It may be semantically contagious. In other words, the same material can co-create and constitute the substance of various compound signifiers, whose meanings can be surprisingly different. In this way there is revealed the ability of theatrical signs to create relations and various correspondences on both levels – the plane of expression and the plane of content, the vehicle and the meaning. The absorption of meanings is therefore a process which is indivisible from the art of the theatre, and indeed reminiscent of a system of connected communication vessels, in which there prevails, however, a variety of pressures – that is why some vessels become empty and others full. In the theatre the semantic value of each active sign increases, grows and accumulates during the development of the events and pictures shown on the stage. This means that other signs, having transferred their meanings, fall out of the game. A handkerchief held by Desdemona early in the play means something else towards its end, when it becomes a sign of an intrigue and the madness of jealousy. If real strawberries appeared as a prop towards the end of the play, their meaning would go far beyond mere fruit. Furthermore, material substances of signs are given time and space attribution by the context in which they appear, which means that they carry this attributed meaning along wherever they are taken or moved. A single sign, a prop, a sound, reveals the ability to clarify space, time and the figure. But first they have to absorb certain semantic inputs.

The use of the same costumes or properties in different performances will be an example not only of the accumulation of meanings by objects but also their (the meanings') transfer from one staging to another. Some time ago this was done in the Stratford (RSC) cycle of Shakespearean chronicles: particular plays were connected not only by history and actors but also by the crown and the royal robe, which accumulated in themselves the meanings of successive monarchs.[11] In such a case the costume becomes a sign not only of a royal cloak in general, the symbol of a monarch's power, but also a sign of itself and of the traces of meaning from previous productions. That is why the former meanings are superimposed like a design or ornament, invisible to the naked eye, on the present ones. On the same principle, an object, being a legible sign of some element or aspect of extra-theatrical reality, introduced onto the stage, will transfer its meanings to other signs of the performance, sometimes to the degree that it creates a new level of readings.[12] Sometimes it is sufficient to introduce a signifying property or costume, recognizable by the spectators, from the reality beyond the stage, in order for it to attract after itself an additional context, to suggest new codes, through which a given performance can be deciphered. For example, in Thomas Middleton's allegorical political drama from 1624, *A Game at Chess*, the actors allegedly obtained for the staging the real clothes worn by the Spanish Ambassador, Count Gondomar, in which was dressed the actor playing the Black Knight. Since Gondomar was a figure universally known and hated in London, his clothes did not only become the costume of one of the figures but also imposed a political context on contemporary readings of the performance.[13] The ancient author Aulus Gellius (2nd century AD) recalls in his work *Attic Nights* (*Noctes Atticae*) the anecdote about a famous Greek actor from the first half of the 4th century BC, Theodoros, who, having lost his son, played Elektra in Sophocles's play; in order to authenticate his acting, he is said to have brought onto the stage an urn with the ashes of his son, so that as Elektra he could weep for Orestes. Those of the spectators who knew what was in the urn deciphered the scene they were watching through the prism of the private life of the actor, in which the tragic episode was imposed on that shown in the theatre.[14]

We have become accustomed to the fact that in a play there are usually leading figures (protagonists) and secondary (episodic) ones. The same applies for objects, some of which are more important (active) than the others. There is no doubt that the main function of the latter is to provide certain information, serving to clarify the characteristics of the main figures, the place and time of the action, and also the missing or unclear links among the threads of the plot. In a word, the semantic capacity of the leading figures grows constantly (we know more and more about them), since they absorb meanings from everything that surrounds them, while the secondary ones fall out of the developing plot like completely unnecessary semantic eggshells. The same applies not only to the figures but also to all other elements of the performance. There is no need to remind the spectator constantly (for example by a costume, a wig or a hat) that the figure he or she is watching is Napoleon. When he arrived on the stage for the first time, various signs had to be employed to define this figure (for example, appropriate make-up, gesture and costume); when the physical characteristics of the actor have been accepted by the audience as a sign of Napoleon, each appearance of this actor upon the stage, even without make-up and costume, will signify that "Napoleon" has made an entrance. This means that the make-up and the costume (the hat!) have passed their meanings to the stage figure's body, enriching its semantic value. There is not even any need to use them constantly (the face and the body of the actor have taken over the meaning of the hat). In other words, the absorption of meanings has such consequences that the homogeneous element of the substance of a compound sign can take over and absorb the meanings of the substantially heterogeneous whole. In this way a simple homogeneous sign becomes a metonymy of a compound heterogeneous sign, and that might be a metonymy of a larger whole. What this means is of great significance to theatre as art: it has an extraordinary potential to broaden and enrich the meaning of the constituent elements of its material substance. It has also a practical dimension: metonymy saves time, matter and energy, it enables us to say more with less effort. This explains why in theatre we are constantly dealing with metonymic multiplication, whereby one simple object or sound or other phenomenon becomes a metonymy of something else, and that, in turn, becomes a metonymy of a larger whole, etc. Please note that theatrical metonymy is always different in the two spheres, divided by the fifth wall. In the real world a simple handkerchief, held by an actor who bulges his cheeks and blows the wind, may become a sign of a sail, and therefore it becomes a metonymy of a real ship, which sails only within the fictional realm; on the stage it is a material prop. Similarly, an actor on the stage is not a metonymy of anything, but as a stage figure he might become a metonymy of an army marching over the stage.

Consequently, in the process of perception, prompted by the actors and shaped by all the signals emanating from the stage, we have the ability to see ("in our mind's eye") the invisible, to take – at least partly – pure fiction as a reality, mental though it is. This results in a form of "dual perception", so typical for theatre: through our senses we see the material substance only (and the relationships between particular elements), and yet are capable of reading the sign, which means that the signals emanating from the stage evoke in us cognitive schemata by which we conceive and understand the world. This is why an actor, even though we see him/her for an hour or so only, becomes a "character" or a "convincing personality" (in plays where that is a factor), tacky furniture becomes the elegant interior of a living-room, and we "see horses where there are none" and have the ability to dress the kings with thoughts only, to use examples that Shakespeare gives us in the Prologue to *Henry V*. In the same play, Chorus tells the spectators in the opening of act 4, to sit down and "Minding true things by what their mock'ries be" (l. 53). Shakespeare wants us to accept the convention, by which the material used is inadequate (hence it is described as mockery), and yet, with the help of imagination, we are able to reconstruct the true shape of "true things". We see an actor and the figure, the costume and the garment, the wig and hair, scenography and a street in a town; we hear verse but know that it is in fact prose.[15] We hear a French translation of *Hamlet*, but we know that French is a sign for English, which, in turn, is a sign of Danish (and in its Renaissance shape). However, the first element in the list exists objectively on the empirical side of the fifth wall, has its material and acoustic substance, the second exists on the second side, only in the mind of the spectator, as a mental reaction to the complex signals emanating from the stage. The first can be photographed, measured, recorded or filmed, the second cannot. Moreover, we do not necessarily have to "see" it – what suffices is our awareness of what the denoted meaning is. We do not have to hear the Danish Hamlet is speaking, but it is important we know it is Danish. We do not have to see Birnam Wood move, but we have to know it does. However, as indicated above, the ultimate meaning derives from the juxtaposition of that awareness to the way, in which the Wood is actually presented on the stage (substantial absence is also a way of presentation). The same rule applies for everything we see and hear on the stage: while remaining itself, an object or a body or a word become a sign of a fictional, hence immaterial, object, a fictional body and fictional utterance. Hence what we hear as a soliloquy or an aside is, from the physical point of view, an utterance of the actor-human being and at the same time a sign of an utterance of a fictional figure that we do not actually hear. However, in all the instances described, the theatricality of that utterance rests on the relationship of one to the other. And that is made possible by the juxtaposition of two models of perceiving reality, the figures' and the spectators', which in fact may be seen as the fundamental marker of theatricality.

During the performance, it is constantly signalled to us by the actors that the figures see the "other side" of the fifth wall only, that is, they do not share our "double view", do not perceive the effect of theatricality in anything around them, they really see the referent, the denoted world which is real to them, and they do not see the substance of the semiotic world of the stage. They see the final shape of the reactions occurring on the stage, but do not recognize the participating components nor the rules and formulas that enabled their formation. They do not see the stage, they do not see the audience, and they (usually) do not reveal any awareness of the theatrical situation they are set into. On the contrary, they see what the spectators do not see. They do not hear what the actors are saying: what they hear are the words uttered in the fictional realm by other fictional figures (where, for instance, the poetic language becomes everyday speech, a soliloquy means silence and an aside does not occur at all).[16] They do not understand topical allusions and do not notice the artistry of the stage-set or of the acting or poetry (the poetic or aesthetic function appears only to the spectators). Similarly, they seem to notice the presence of actors' bodies, but they do not recognize these bodies as belonging to actors, but as belonging to Macbeth, King Lear or Uncle Vanya, and they hear seemingly the same words, but these words mean something different to them and to us, the spectators. This discrepancy is the essential feature of theatre as art.

Generating meaning

Generally, it may be said that on the stage, the meaning of particular material components depends on at least seven conditions: (1) the intrinsic physical characteristics of a given element or signifier (fabric, colour, shape, tone, pitch, brightness, etc.); (2) its relationship to the denoted object, body or phenomenon; we have to remember that the relationship might not at all be obvious: iconic similarity of stage signifiers to the signified is often very vague, if identifiable at all; (3) the network of relationships between particular components of a given compound signifier (e.g. word, gesture, movement, music, light, which constitute a single blend); (4) the relationships between the whole sequence of compound signifiers (their clusters) constituting the performance; (5) the intertextual relationships with the dramatic text and other cultural, political or ideological texts; (6) the individual characteristics of the spectator; and on (7) the external context, political, social, cultural and economic, including the biographical and personality details of the author, director or actor.

Of the seven conditions enumerated above, the first five are the objective and invariant features of the stage "text", whereas the latter two change in time and for that reason are variable, and additionally are culturally determined. Theory of theatre predominantly concentrates on the first five, which means that not necessarily neglecting the context, it concentrates basically on *how* the meaning is created, what are the rules and formulas, what stands behind the particular selection of substances used in particular production, what is the internal mechanism that has the ability to generate meaning in a way peculiar and distinct to the medium.[17] Theory is not that much interested in the meaning generated, as in the ways it is generated. Also, in cognitive approaches, we are interested in *how* the meaning affects the recipient. Sociological or ideological approaches, dominating today, concentrate on the sixth and seventh conditions and tell us what the social function and meaning was at a given time and place, what the reader/spectator response was and what were the social or political implications or repercussions of the performance.

The role of context cannot be overestimated. It may play a decisive role in the creation of meanings at a given time and a given space, for instance, when a stage performance becomes an element of a sequence of different (usually) theatrical, paratheatrical or ritual texts or events (such as parades, firework displays, demonstrations, state or religious ceremonies), creating a kind of "supertext" and performed in spaces other than performance ones. It must be noticed that the inclusion of a performance in the supertext, and therefore its transformation into a new structure, may not have been foreseen by its creators (as, for instance, the inclusion of a Shakespearean play into a sequence of court masques and rituals). Secondly, a meaning peculiar to the "present occasion" appears when a performance takes place and is deciphered in the context of a clearly defined time and space, different from the original foreseen by its creators, and therefore employing codes which are not appropriate to it, but are external. The combination of these codes is generally not consistent with the integral structure of the (original) performance, but is creative exploration (or ideological conditioning) that can lead to the conscious creation of a spectacle in such a way that it creates an intentional and congruent meaning, also in a strictly defined, although transitory, context of time and space (social, political and so on). This same stage event, presented at another time and in another place, will no longer be a piece of political theatre. The well-known example of the staging of *Richard II* on the eve of the Essex rebellion may serve as an example of the spectacular significance gained by the text on that particular occasion, meaning that could not be repeated on any other.

In the first of the examples described above, when incorporated into a larger sequence of events, a performance loses the independence of an artistic work, becoming an element of a greater text and its structures; in the second – in a given overimposed temporal or cultural context – in a concrete reception it loses this independence as a consequence (often tendentious) of the inclusion of (in)appropriate codes of deciphering, through which it becomes a commentary, a confirmation or a confrontation of texts and codes which are external to it (for example, current events or political, social, religious or economic tensions). In both cases the performance becomes a text referring to other spaces and texts, which are known to the recipient, and is deciphered in relation to these texts. In other words, it draws the spectators' attention away from the stage to reality outside theatre, and thus becomes "a communiqué within a communiqué" (as in Aesopic literature). In partly losing the independence of a work of theatrical art, it becomes a piece of political or pedagogical theatre (and therefore most often a work of journalism or propaganda, sometimes of a satirical nature, performing the function of a "distorting mirror"). It may also provide a commentary on the author's, director's or the actor's life and deeds. It must be noticed, therefore, that meaning is not necessarily a permanent feature of a given stage performance, but is, at least partly, imposed from outside by the varying method of its realization, the context and the specific process of deciphering.

It may be added that in recent decades many productions show a tendency to signal the precise context they want to be read through; this is done through verbal and nonverbal signals evoking, metonymically, iconographically, or through quotations, all sorts of ideological texts that are brought into a given production as a sort of immaterial spatial context or frame that serves the function of an explication. In this sense, a prostitute's costume which Kate (*The Taming of the Shrew*) is forced to wear is not necessarily a sign of a prostitute's dress in the fictional world, or in the Renaissance, but a sign of the role of a woman in patriarchal society and culture of the present time (as in Krzysztof Warlikowski's 2003 Warsaw production). As an iconographic symbol, the dress transfers its ideological message onto other elements of the production, which now may be read through the prism of other ideological texts dealing with feminism, gender, patriarchy and the like, providing an intellectual and cultural context which justifies such a reading.

I have already indicated that everything we see and hear on the stage comprises signalling matter that has been selected and combined *ad hoc* for a given production. Beside the axis of selection, traditionally differentiated in language studies, we can talk here of the process of modelling, i.e. of creating the shape and composition of the material substance of a complex or compound signifier, being a significant element of the creative process in theatre (such as, for instance, casting or designing the stage set or the costumes). We could add, after Roman Jakobson, that selection is directed by the principle of equivalence, similarity or difference (the so-called *paradigmatic axis*), while the axis of combination is governed by the principle of dependency (the so-called *syntagmatic axis*). Both of these rules (selection and combination) determine the complicated network of relationships between particular components, which is the most important factor in the theatre's ability to generate meaning and to evoke the aesthetic function. The rule of combination or contiguity relies on the fact that all sorts of components visible and audible on the stage at the same time form divers spatial configurations and compositions, which are not accidental; the rule of selection points to various relationships or equivalencies between elements of the performance that are not contiguous in spatial or temporal terms (a prop, a word, a sound or a gesture in act three may create an equivalence with another prop, a word, a sound or a gesture from act one). As Anton Chekhov put it, if a cannon appears in act one, it is likely to fire in act three. This explains why in theatre everything enters into close or more distant relations with everything else, even though particular components do not appear on the stage together (this is what we mean when we talk about a theatrical production as a structure: it implies a network of equivalences, functions and relationships). The latter effect is possible as a mental cognitive process by which the spectators perceive the message.

Thus, what is essential to theatre is the signalling matter that acquires a new function to mean something it is not (i.e. something it is acknowledged to be in the fictional world), a human factor, i.e. the actors, who, through their utterances and behaviour describe to us what the acquired meaning is, and the spectator who is capable of establishing a relationship between the two, to blend them in complex cognitive processes. As in an alchemist's dream, on the stage everything may become gold if it glitters; moreover, it may be gold even if it does not glitter. It is worth noting that the rule of combination in theatre is of particular aesthetic importance and is conspicuously different from that in literature or language. The contiguity of particular elements in theatre is not only linear but also spatial, for they appear on the stage simultaneously, creating three-dimensional "pictures" or pictograms. Thus, the rule of combination in theatre governs the composition of an individual picture (scene), whereas the rule of selection points to the equivalences between this and other pictures (scenes), of which the whole performance is constructed.

Thus, it should be noted that in theatre, the material selection, modelling and combination of stage signifiers is of major artistic and semantic significance. Hamlet's physical appearance, his costume, voice pitch, make-up and wig are not irrelevant to the meaning and artistry of the whole. In Eimuntas Nekrošius's production, he wears a shirt that is made of material that dissolves under the shower the actor at one point takes on the stage, while delivering the "To be or not to be" soliloquy: in this way the director creates a visual metaphor of the protagonist's torn emotions or, perhaps, foreshadowing his later "changing his skin". The metaphor may, of course, be created verbally, or through some other visual effects, but in every case, the meaning and the aesthetic function will derive from the relationship of the attributes of the fictional being or element (the figure of Hamlet who puts his "antic disposition" on) to the material substance of the sign(s) that denote it and to their modelling (here: the body of the actor, his gestures, movements, costume, water, lights, music, etc.). Please note that with this visual context, the shirt dissolving under the shower, all the words of the soliloquy acquire meanings unique to this particular production. It should therefore be noted that the particular aesthetic effect cannot be translated into any other "language", and remains unique.

Crumpled paper becomes Yorick's skull, as in John Blondell's *Hamlet* produced by Lit Moon Theatre Company from Santa Barbara (seen by the author during the annual International Shakespeare Festival in Gdańsk, Poland, in 2004). And that paper appears in the context of the paper stage set, and thus metaphorically may be related to the substance of the world, equivalent to biblical dust; it may be also related to the words Hamlet is reading in a book (thus the book, made of paper, becomes a metaphor of the world). But paper as such does not contain any of those meanings, it has no direct links with the world of Hamlet nor with a biblical code nor with the contents of the book you are reading at this very moment. But in that particular production paper gained this meaning through a network of relationships with other elements of the performance, and it also gained the ability to convey various cultural codes.[18] In other productions of *Hamlet* this particular sign function will not be applied, and, consequently, the same meanings will not be generated (but others will). Even if the material dust is used as a stage metaphor, as is the case in Nekrošius's production, mentioned above: in it, the director employs dust, an obvious symbol of both the act of creation and of death and transitoriness of life, as one of the four elements – with the remaining three being water, fire and air, to create a fascinating network of relationships between words and the material substance of that particular production.

Sunflowers are not contained in the word, nor are they present in the Van Gogh painting. In the latter, what is presented with paint in two dimensions may only be similar to the three-dimensional sunflowers in the real world. On the stage, we can imagine an actor's costume, which resembles the sunflower, or a three-dimensional plant made of papier mâché. The similarity enables us to call this kind of a sign iconic. In theatre, however, the sunflower does not have to appear as a material sign similar to the real thing, for it does not have to appear as a material object at all – it can be denoted by the actor's word, gesture or gaze and a combination of the three. We notice that in the material substance of the sign there is no similarity whatsoever to the denoted object, but we notice that its meaning derives from the sign's contiguity with another object or somebody's gestures and movements, or from the fact that it appears in a chain of events that is based on cause and effect. This is the case of indexical signs that theatre rests on. To use a contrary example, when on the stage we see the real sunflower, it may be deprived of its original meaning: we may be led to understand from the way it is perceived by the fictional figures that it is a model of the globe, a warrior's shield, a mirror or a plate full of soup. What we see on the stage (the sunflower) bears no relationship to the plate of soup (with the possible exception of common quality, being the roundness). The relationship is established by the behaviour and gestures of an actor who eats out of this "plate", or looks at his/her image reflected in a "mirror"; thus, his/her gestures and facial expression, the movement of the jaw, all point to the different meaning and function of the object beyond the fifth wall. This we may treat just as a convention, which conveys unusual meanings and, at the same time, uncovers the rules of theatre, but the said sunflower may indeed create all sorts of equivalences with other elements of the stage set, the design and colouring of the costumes, with the language and the images and the metaphors it articulates. In this way, the simple prop may gain new allegorical or metaphoric meanings that go far beyond the semantics of the plant. Metaphorically it may be said that the plant or fruit ripens semantically during the whole time of the performance.

Thus, let me repeat, the essential meanings in theatre are created through the clash and confrontation between at least two models of perceiving reality (the figures' and the audience's). Without someone signalling to us (verbally or through gestures) that X is Y or that the change of purple to green means a change of space, and that an empty glass is full of exquisite wine, we would not be able to create a coherent model of fictional world in our minds. We would not grasp the whole chemical reaction and its outcome. However, in order to signal all that an actor has to reveal the figure's awareness of its "I" and of time and space it is set in. Without that awareness (which has to be linked to *deixis* or the deictic axis) it is impossible to signal a different model of perceiving reality, its temporal and spatial features, hence no theatre is possible.[19] This is why objects and animals without human support cannot create fictional temporal and spatial dimensions. As noticed by others, the play and interchange of *deixis* is of particular importance in theatre.[20] In the temporal sense, linguistic communication or events on the stage always acquire meanings of the historical reconstruction of utterances or events, which in some assumed reality beyond the stage were made or happened in a more or less defined past. It could also be the future – then the dialogues and events are not an assumed reconstruction but a projection (in rhetorical terms, their *proleptic* function will be revealed). Accordingly, all linguistic utterances on the stage are characterized by an apparent internal contradiction: that which in a grammatical sense creates the impression of the present, the *hic et nunc* of the spectators, is in reality a re-creation of something which somewhere at some time has already happened or will happen.

It seems that in the success of the process of creating fictional time and space lies the essential task of the theatre as art. How does one create these fictional structures? Well, since this is a physical impossibility, the only way out is through a sort of agreement between the performers and the spectators as to the rules of the game, reminiscent of the way children play. The method of bridging the gap between reality and fictionality is through convention. Thus, in my understanding, *the basic formula of the theatre occurs when by mutual agreement at least one person (the performer) pretends before another (the spectator) that what belongs to the past tense (or future) of the spectator is the former's temporal present; in most cases the one who pretends will also pretend to be someone else and at another location.* This implies a rigid division of the stage message into two interrelated spheres, governed by different and incongruous times, the one sphere that is visible and audible to the spectators (phenomenological sphere), and the other that is only implied or denoted by the first one, a sphere that is immaterial hence fictional. However, each of the spheres is perceived as reality by those who inhabit them. This also means that the spectator can at least partly experience, as cognitive science would have it, the actions and emotions of a fictional figure (as enacted or described by the actor).

The contradiction between different models of perceiving reality is deepened by the fact that time flows with a different tempo in each of the spheres, undermining what we know about physics and biology;[21] moreover, in the fictional realm it reveals unique features, such as its elasticity and flexibility or ability to slow down, accelerate or even to stop completely. Please note that it all rests on the ability to model the fictional time, and cannot be achieved by humans in their real biological time in any other way than through playing games or acting: in real life one cannot speed real time up, or slow it down, or suspend it. One cannot make the present elastic. This is why a temporal ellipsis, although possible within the fictional time, is generally avoided in theatre (it would violate the duality of the present time, which ought to be maintained at least within a given scene). One cannot employ ellipsis in the real present time, which cannot make "jumps", or be edited. This is why, unlike in film, in theatre employing ellipsis in the fictional stream (verbally, for instance) destroys the duality and coalescence of the theatrical present time. Theatrical time may flow with a different tempo, but cannot be composed of segments or smaller units. And it must be noted that the difference of the two streams of time is experiential: according to cognitive psychology humans can in fact experience time, the real time they live in, and this may be contrasted with the awareness of the fictional time, which is an intellectual construct. However, it seems that in theatre illusion might be the factor that enables the spectators to experience the fictional time too. One may doubt if the synchrony of temporal experience is possible, but its alternating sequence is plausible, and requires further psychological scrutiny.

Thus, as explained above, the creation of different time structures and the juxtaposition of different models of perceiving reality are the essential features of theatre as art. The two incongruous models of perceiving reality can logically co-exist only if separated by time. It is the fifth wall that creates the invisible divide between the two streams of time and the two present times, and between the two spaces, the real and the fictional one. This becomes the marker of theatricality. The fictional realm, including the figures, belongs to a different ontology and different time and therefore cannot communicate with the reality of the spectators. When it does, i.e. when the actors behave as if the figures were aware of the presence of the audience, addressing the spectators directly, or were capable of perceiving elements of our reality, it would violate the basic agreement and rule of the theatre – the rule of accessibility: in its material substance the world of the stage is accessible to us, spectators, whereas our world remains inaccessible to the stage figures. This creates the afore-mentioned dissonance between what we see and what we do not "really" see, between our, the audience's, full awareness of being present in the theatre and that which the figures do and do not do, see and do not see. This is the consequence of the invisible wall that separates the material world of the stage from the immaterial realm that the former denotes, it separates different laws of physics that operate in each of the spheres. This is precisely what I call the *fifth wall*. The rigid division of the two worlds may of course be weakened or even suspended, whenever the spectator adopts, at least partly, the model of perceiving reality of the figures, which may result in the spectator's strengthened experience of the figure's actions, utterances and emotions.

Kaleidoscope

I am therefore proposing to look at the theatre as a unique form of human communication, in the process of which one side, primarily the playwright, the actor and the director, tries to communicate something (the scenic communiqué) to the other side, primarily the spectators (the secondary addressee might be the media). The director attempts to establish chemistry between his/her work and the members of the audience, which may also engage a certain dose of shared experience in spectators and the figures. In order to achieve that he/she projects a message onto the stage, a message contrived in signs that are not purely linguistic, but predominantly visual or iconic. The signalling matter, of which the material vehicles of the signs are composed, is arranged both spatially and in a linear succession. Consequently, every material component reveals a dual orientation – as part of the spatial arrangement, where it appears simultaneously with other components, and as part of a sequence, where it enters into different configurations and relationships. A prop in an actor's hand is part of the composition of a particular visual and three-dimensional unit of a scene, a stage "picture", and at the same time may create all sorts of relationships with the words uttered by the actor, which, in turn, create meanings in their linear arrangements. On top of that, additional meanings are created through the equivalences created along the paradigmatic axis, by which a prop and a word in scene one, may reappear in a different context in scene nine. The word and the prop bring these events together, and their relationship becomes significant. This indeed is reminiscent of a three-dimensional kaleidoscope. The spatial arrangement does not require time to appear, the linear one does. This means that every object and phenomenon on the stage, in Roman Jakobson's terms, is both a spatial and temporal sign. Consequently, every word, material element or phenomenon enter into relationships with everything that occurs simultaneously within the space of the stage, and at the same time into different relationships with every word, object and phenomenon that precedes or follows it. Please note that seen from this perspective, individual stage pictures, taken as a whole, can evolve only in a linear manner, which means that they can create additional meanings as they shift in time.

We can therefore, traditionally, consider theatre in the category of a communicative act, an utterance which is transmitted through at least two axes or channels, direct and indirect – with the former not being perceived by the fictional stage figures. Shakespeare was fully aware of this phenomenon and although he never expressed it in exactly similar terms to what I am attempting in this book, many scenes in his plays, such as the play-within-a-play in *A Midsummer Night's Dream*, show us the mechanism of the theatre as a system of distinctive rules. It follows that communication in theatre does not take the form of a usual conversation or dispute, but is of a different kind, based on the one hand, on the artistic spatial and sequential modelling of the performance text, which may be seen as a unique exteriorization of the creative mind(s), and, on the other, on the spectator's cognitive process of perception. As I have already mentioned, it is an ensuing dialogue between the speaking picture and the silent brain. The spectators are expected to deconstruct the specific modelling of the performance text, and through that uncover the cognitive space that lies behind it, leading (or not) to the understanding and justification of the materials used and the ways in which the text is constructed. Without an active part on the side of spectator, no act of communication can take place. Thus, the chemistry of the theatre has to be seen as an act of specific communication, and not just a piece of information or mere event. This means that the reactions are not taking place for their own sake, but are directed towards conveying stimuli and information from the sender (whoever that is) to the recipient who is expected to take an active part in the act. Please note that the activity of the spectator does not rely on his/her immediate or physical interaction with the performer. Someone wants to tell us, the spectators, that a certain way of life or worldview is superior to any other, or that we should cherish certain values and reject others, and in order to achieve that goal, to construct the message, uses very unusual means and substances. The creator also wants us to notice *how* his/her message is composed, what is the material substance and its modelling, what are the relationships between the components engaged in stage reactions, and what are the rules that enabled the appearance of particular blends. The performance text is, after all, about its creators, and not only about some fictional figures' vicissitudes.

This also means that at least part of the information provided by the director, actor or scenographer, people involved in the creation of a given production, is oriented towards itself, towards the signifiers and their modelling. Theatre talks about itself, shamelessly uncovering its anatomy and substantial "artificiality". This means that all theatre signs reveal a double orientation: on the one hand they are oriented towards the fictional realm, which they denote or describe, on the other, towards their own material substance and the rules of modelling.[22] They are self-referential. Substance is (part of) the meaning, a fact often neglected in scholarship. And it is the signifiers that react on the stage producing the signified, which are only implied or denoted. In this way, the chemistry of the theatre blends the material substances visible and audible on the stage, engaged in all sorts of reactions, with the mental space of the spectator where the effect of those reactions appears. This provides the basis for the final reaction to take place, which is the blending of the three spaces, the generic (authorial), the phenomenal and the signified; thus, let me repeat, the ultimate meaning in theatre is not simply referential, the immediate signified of the stage signifiers. The perception of a stage "picture" requires at least three phases: first we recognize its material component as equipped with a sign function, i.e. with meaning that goes beyond the literal or the obvious, then we deduce the denoted fictional referent, and then we relate it back to the substance and modelling of the signifier(s). From what we hear and see it appears that a given scene takes place in a garden. This is the immediate signified of the stage signifiers, which may, but do not have to, reveal the physical features similar to those of a garden. But, to give just one variant case of the example given, what we see is a piece of a flat painted canvas with fruit trees and a green carpet that pretends to be a lawn. We are thus led to understand that a convention has been employed, by which the garden is represented by its cheap substitute, a metonymy if you like. The inadequacy of the material used and its modelling become a stylistic trait or marker, by which we can adjust our perception of the performance, based on the assumption that it does not pretend to be "real"; on the contrary, it unveils and foregrounds its conventionality. All of these aspects may also influence the spectator's experience of the performance.

Thus, theatre is not a simple act of human communication know to us from everyday life. A significant part of the stage communiqué is conveyed indirectly, as if the spectators were not present, and someone else was the true addressee of the message, which puts the spectators in the role of eaves-droppers. Also, there is little opportunity left for visible or audible interac-tion, which is limited to its minimum, such as the spectators' approval of disapproval, signs of understanding or boredom (leaving aside interventions of angry or militant spectators, theatre riots and other incidents known from theatre history). And, anyway, the performers seem not to notice these responses. The actual interaction is predominantly mental and experiential. The situation in the theatre reminds rather of an apparatus where a three-dimensional, illustrated book is presented to the readers, the pages of which are also inhabited by live human beings. We, the spectators, "read" the text as a sort of three-dimensional cryptogram, and the actual act of communica-tion is a cognitive act. The cryptogram is spatial and consequently provides bundles of information simultaneously, and for this reason we have to learn how to identify, select and read those signals, which, out of their multitude, react with each other at the same time. This is not an easy process, but theatre is art and will provide works or events, which at least partly will be oriented towards themselves, by which they reveal their inner features and explain and justify the shape (the "form") they are given. This allows us to speak about theatre as a self-explanatory medium, as opposed to much of today's art, which cannot explain itself without a theoretical or critical or ideological explication. This is why it is not obligatory to know the "grammar" of theatre beforehand.[23] Only in the case of highly conventional theatres it is obligatory that the spectator knows the formulas beforehand; also in the case of some postmodern or postdramatic productions it is necessary for the spectator to be aware of an ideological credo that lies behind them. Without it, the performance often hardly makes any sense.

Thus, in theatre the chemistry occurs between the invisible author, the actor, and the mind of the spectator, the two or three spaces or retorts. The act of communication appears in the formation of the closed circuit or feedback between the receptive mind and the phenomenology of the stage. This may remind one of forming a cognitive crystal in the mental space of the recipient's mind, with the stage providing the saturated solution and the catalyst, and the mind of the spectator being the vessel (context) for the process of crystallization. As in chemistry, theatrical compounds are not simple mixtures of components, but become inseparable as elements of structural units of a larger whole. They are in fact a network of relationships. Let me repeat: in theatre it is not only the referential meaning that counts; we all know what *Hamlet* is about. What is of crucial importance is the substance, modelling and the network of relationships between particular components of a given production. The plot, understood as a sequence of events, may be known, but what really counts is how it is presented. Peter Brook's Hamlet and Eimuntas Nekrošius's Hamlet might be uttering the same words, but these words mean something substantially different. The language component seems the same, but it enters into reactions with all sorts of material substances, selected and combined in a manner unique for a given production, which results in original and unique compounds or blends. Words create unique material amalgams, blends, by which they become predominantly iconic and create meanings and aesthetics peculiar for a given production. This is why it makes sense to see new productions of, say, *Hamlet*, even if we know the plot, because the cast, music, lights and objects, their modelling and their relationships will inevitably be different from all the previous ones, and will therefore mean something else – also in the aesthetic sense. To relate this meaning to the alleged meaning of Shakespeare's *Hamlet* in a qualifying manner, or as being "faithful" or not, is an absurd procedure.[24]

All of this means that the theatrical interaction is of a special kind, because it combines the phenomenology of the stage and the mental processes, i.e. conceptual blending, occurring in the spectators. The final success of that act of communication or chemistry between the two sides involved, the co-operation of the two retorts, depends on many factors, but, as hinted above, the initial and most important one is the agreement between people responsible for production and the recipient (as a rule, both sides have to agree to take part in an act of communication). What is the major issue of the agreement? Well, it is the readiness of one side, being the performers, to create "different conditions" of being for various preselected components, which enables them to take part in the prearranged reactions on the stage, so that the fact may be recognized by the spectators and, consequently, the space and time of the performance may be distinguished from those in the non-theatre world. In order to achieve that, we, the spectators, simply agree to watch and listen to some people who, within a given space, amid sundry objects, light and music, behave and talk strangely for two or three hours, and we agree to treat this not as the strange behaviour and utterances of lunatics, who have lost the sense of identity, space and time, and do not realize where they are, and do not notice our presence, but as "characters" of an encoded message unfolding in front of us, by which someone not visible on the stage in an unusual way wants to tell us something about him/herself or about the world. The people on the stage, on the other hand, agree to behave strangely for two or three hours, pretending as if they were not themselves, and not noticing the situation they are in (i.e. a theatrical performance), and not realizing that they and the objects around them are the substances of the stage text that someone is watching. They agree to utter words that are not theirs, and to pretend to be someone else, living at a different time. They speak (and at what length!) now in verse, now in prose; they sing, dance and pretend to be living in a different era; they have made-up faces and wigs; they die, marry, betray each other, fight, commit crimes or indulge in drunkenness. And it does not really matter whether they do this as a hobby or as a professional undertaking, whether they are paid or not. Anyway, without the agreement, a perception of a theatrical performance is likely to lead to misunderstandings.

Conceptual blending

The dialogic or communicative character of a theatre performance may in fact be explained in terms of *conceptual blending theory*. It is possible because the stage space, filled with its signalling matter, selected and modelled distinctively for each production, may be seen as a final visual and verbal "description" or, rather, an exteriorization of conceptual blending in the mind(s) of the director, playwright and other people responsible for the final artistic shape of the performance. It is a material exteriorization of the creators' individual blend, and includes the mapping of at least two major blended spaces, each flowing within a distinct stream of time, the real and the fictional. Thus, the spectator "sees" (reads) both worlds, one real (metonymic, inadequate and fragmentary in its material nature), the other, implied, hence immaterial and conventional. These are the general input spaces that require further complex conceptual blending. Within the general spaces there appears an evolving sequence of innumerable minor input blends, which contribute to the cognitive processes in the mind of spectator. Owing to the spectator's standing point, he/she alone has a complete vision of the performance and hence is able to grasp the overarching structure of the artistic work, and is also able to experience and interpret it (the actor has no such ability). What is more, as I have said, perception of the performance is possible thanks to the imaginary structure of the fictional world (its model) created in the mind of the spectator. It is also possible owing to the constant play of various blends, which is done below the level of consciousness. What on the stage is a metonymy is "sewn in" and integrated ("synthesized") in the process of perception of the performance, the process of cognitive blending, and the mental structure, the network of categorizations, created can be transferred back on to the signalling material of the work, including the actors (in reading a drama there is obviously no such possibility).[25] Thus, the input spaces, the stage with its signalling matter, and the fictional realm, are juxtaposed and blended together in the conceptual blending process in the mind of the spectator. And in this sense the spectator becomes a partner in the act of communication, and a (co)creator of the meanings and significance of the performance, including also the figures. It is he/she who "translates" the semiotic structures into semantic ones, at the same time recognizing the aesthetic factor. The record in his/her memory (sometimes derived from earlier productions or from reading a given work, or even from subjective mental images), constantly changing under the influence of new stimuli of observable relations or blends, is continually remodelled by the actors and other components of the production. This creates a closed circle of a theatrical act of communication: the stage signals evoke cognitive responses in the spectator, and the latter enable and influence the act of perception, which makes the spectator to look at whatever is happening or is being said on the stage with a growing awareness and acuteness, which allows him/her to grasp the complex networks of relationships, constantly developing on the stage. The pleasure of understanding is part of communication in theatre: the common language has been found. Moreover, as discussed in Chapter 1, mirror neurons are activated and allow the spectators access to the emotions of the stage figures. The latter phenomenon is sometimes called simulation, which, in theatre, rests on the ability of embodied emotions to produce corresponding emotional states in the spectators.

What this means is of great relevance to theatre studies, for as Bruce McConachie observed, the new discoveries in science have provided evidence "that it is *spectators* who mirror the motor actions of those on the stage; cognitive imitation is a crucial part of spectatorship … Mirror neurons do not invalidate Aristotelian mimesis, but if we are interested in audience response from a scientific point of view, the mode of imitation triggered by theses neutrons (and their consequences) should be part of our explanation."[26] It is clear that further research will probably cover areas of investigation previously considered not empirically verifiable (such as language or metaphor comprehension). It is not enough to say that activated mirror neurons simulate the experience of corresponding states of emotions in spectators: according to the theory, this also happens outside theatre, on occasions that do not have to be artistic at all. But to my knowledge, neuroscience has so far not explained the unique experience of artistic texts and of theatre in particular. It has not explained the phenomenon of simulation when the observed object is fictional, as is the case in the theatre: naturally, we see the actor at work, but we experience the enacted person's emotions and feelings, not the actor's handling of the role, nor of his/her technical skill, nor of his/her emotions (of which we, the spectators, practically know nothing).

The cognitive process described may be treated as a scenic metaphoric visualization of a *conceptual integration network*, which is one of the phases in the sequence of the blending process. Moreover, the spectator is expected to "read" the text projected on to the stage, and through its deconstruction to retrieve the input spaces, by which the rules of encoding or their cross-space mapping will be uncovered, will become clear and therefore coherent and appropriate. Through this the work of art will unveil and explain itself. It will justify its appearance in the substance selected and shape given. The *matching and counterpart connections* are visually and orally presented on the stage. At the same time, apart from identifying the input space, the spectator is expected to find a generic space, which shows the common features of particular inputs. Generic space is related to specific blended spaces: "Blends contain generic structure captured in the generic space but also contain more specific structure, and they can contain structure that is impossible for the inputs"[27] such as, for instance, the co-existence of an actor-human being and the fictional figure. Again, the stage picture, with human bodies, objects, light, movement, acting, props, music, etc. may be seen as multiple blends, which constitute the possible relationships within the substance of the stage text, and which the spectators are expected to deconstruct in order to recognize the input spaces, a process which in the case of theatre cannot be automatic. Through this the stage text justifies its own existence in the shape given, and provides the spectator with the joy of understanding. Not only a common language has been found, but now the spectator understands why in order to say this or that the director has created the language for this particular production out of specific and particular components, and why he/she has sculpted them in an *ad hoc* manner. Through this process, the spectator may retrieve the formula that lies behind the reactions of various components on the stage, may reconstruct the *generic spaces* that display the shared features of the *input spaces*. Nonartistic texts can be articulated in all sorts of ways, and the medium used does not add significantly to their meaning; in the case of artistic works, theatre in particular, the situation is totally different: the medium conveys information, so does the "form" in which it is conveyed. It is substance-specific.

We understand that the fictional figure's reactions to and utterances concerning the world around it are the effect of its perception and cognition of that world. We see the material spaces that seemingly function as the input spaces in the cognitive process of both the sides involved, the figures and the spectators. However, we notice that the outcome of this process is different and incongruous, sometimes strikingly so. This forces us to deconstruct the figures' utterances and behaviour in such a way as to find out their true cause, i.e. to deduct the implied fictional spaces that may have to the results we see and hear. The sunflower we see ceases to be a sunflower on the other side of the fifth wall, and becomes a plate of soup (I have seen a production, where a bicycle wheel served that function). We imagine a plate of soup, or simply acknowledge the fact that it exists in the fictional realm, and that alone justifies the otherwise absurd behaviour and utterances of the actor. Thus, part of understanding in theatre relies on the spectator's ability to deduct the causal input spaces that justify the figure's description of its world. It is not only the imagined plate of soup that is the intended meaning of the scene: what counts is how this meaning is created, what substances are used and how they are modelled. In cognitive science it is even more important how the blending process affects the body and mind of the recipient. It seems that the near future belongs to this area of investigation.

3
Chemical Reactions, or Blending the Components

Compounds

As in chemistry, also in theatre, there are laws and formulas that govern all sorts of reactions taking place on the stage, and these laws enable the director to create meaningful performances, structures composed of many different substances, of which language is only one. However, in the science of chemistry the compounds and elements, taken at their "face" value, mean precisely what they are and nothing else, unless they are part of a different, external activity that involves conveying information (such as a police investigation, archaeological enquiry or geological search). In those cases chemical substances have the capability of telling us a lot about external features or circumstances and the time of incidents and findings that are connected with a world that is not necessarily only chemical. They reveal this feature owing to their ability to show contiguity or other relationships with the world outside; for instance, they may appear as significant elements in a cause and effect chain of events (poison discovered in a glass of sherry being the cause of somebody's death, or a substance that added to petrol choked the engine of an aircraft causing the death of a politician). Thus, the specific and singular context assigns to a chemical compound additional meaning, which enables us to describe and explain the world around us, and various plots in which humans are entangled. However, the meaning is not intrinsic to the chemical substance. Nitro-glycerine may be used as an explosive and as medication in heart diseases. In both cases the composition and the formula of nitro-glycerine are exactly the same. Large doses of a number of medicines are lethal. But death is not the intrinsic or intentional referential "meaning" of any of these substances; it is neither signalled by any of its constituting elements nor by their structure or composition. The formula does not carry that sort of information. Neither does the physical appearance of nitro-glycerine contain any information, apart from its being a liquid substance (as medication it is taken in pill or spray form, which signals to us that we are not to use it as an explosive). The system of chemistry is not semiotic (elements and compounds do not carry a sign function and consequently are not signs of what they are not), and the meaning can only be added or assigned. Moreover, in real life, nitro-glycerine, to exploit the example further, does not enter into a network of relationships with all elements of what we call life; on the stage it does, that is, if nitro-glycerine is mentioned verbally or appears as a stage prop, it enters into relationships, closer or more distant, with everything else that is said and what phenomenologically appears during the whole performance. This is exactly the feature of every single signifier that constitutes a stage production. Thus every component and reaction finds further reverberations in the stage retort. We are dealing with a structure of interrelated substances and reactions, and the ultimate result is the meaning of the whole, as it appears in the cognitive processes in the mental retorts of the spectators.

In the complex chemistry of the theatre all sorts of substances are used, which in themselves mean or do not mean anything in the outside world, and during the *semiosis* (with time being the binder) they undergo a process by which a new meaning is added to them or their intrinsic one is altered or almost totally annulled. In the same way a meaningless sound, a phoneme of a language, when combined according to certain rules, with other sounds or phonemes, may constitute a word, and a sequence of words will form a sentence, and a sequence of sentences will create an article or a novel. The selection and combination of sounds will first be governed by the rules of the language, but then may fall under the rule of a different set of laws of a different system, which we call literature. And that literary text, in our case being a piece of drama, may fall under the rule of yet another medium we call theatre. So, there appears a hierarchy of rules, under which all elements of the performance are organized into hierarchies of interdependent structures. This means that any whole belonging to the external world, when incorporated into the performance, becomes only a part of larger unity. Thus, an independent work of art, such as a painting, a piece of music, sculpture or a poem, loses its independence or autonomy, and becomes entangled into a larger structure, by which it loses its original features (also its systemic ones) and through becoming an element of the whole performance gains new features and functions. A painting on the stage, even if original, becomes an element of the whole set, and may create totally new relationships (hence meanings) with objects, music, light and words. A sonnet, when recited on the stage, ceases to be just a sonnet, it becomes a sonnet recited on the stage. Owing to that factor it enters into all sorts of relationships with other verbal and non-verbal components, generating meanings unforeseen by the original sonneteer. It will inevitably create a compound signifier with the actor's face, his/her make-up, wig or costume. It may but does not have to be recognized by the figures as a sonnet.[1]

However, to determine what exactly enters into stage reactions with what, is at times not at all obvious. It is rather certain that it does not happen at random, but according to an overall plan, which is part of directorship. But sometimes, and in recent times more often than not, the spectator is at a loss and does not know how to make sense of what is being presented and said on the stage. Quite often the latter contradicts fundamental preconceptions that a spectator might have about what theatre or art is. What is then needed is a sort of prompt-book, which would clarify the rules behind what is seen and heard, by which what seems chaotic or inscrutable becomes clear and justifiable. Outside theatre there are chemical laws which enable certain elements to merge or to disintegrate under certain circumstances. There are laws that limit an element's or compound's participation in certain reactions, or, on the contrary, enforce an unforeseen reaction. Sometimes a third party or component is needed for the reaction to take place, and that is called a catalyst. In the course of chemical reactions, elements and compounds change their shape, colour and state (solid, liquid or gas), usually beyond recognition. The result of these reactions usually does not reveal any physical similarities to the features of the participating components; it does not share their characteristics and properties. We cannot see sulphur in sulphuric acid, or hydrogen in water; neither can we see all the metals in an alloy. We do not see chlorine, being a gas, nor sodium, being a metal, in salt (sodium chloride).

The problem, however, might be that the reactions of visible and audible substances on the stage will not produce a clear signified: we do not understand the formula, we cannot grasp the final effect of the scenic reactions (although the lack of coherence may be intended). The play, as it unfolds in front of our ears and eyes is impenetrable: although in many cases we know the plot, we do not understand the way it is modelled on the stage, we do not know how it is going to be articulated in later scenes. This evokes curiosity, suspense, nausea, or, simply, cognitive tension, which forces us to look for the explanation of what we see and hear, to find out the formula. We see the philosopher's stone on the stage, or, rather, the ingredients that led to its creation, but we like to know how it was achieved, to comprehend the choice of ingredients and the formula that led to the stone's appearance. Eventually, this may lead to the spectator's understanding (or, lack of understanding) of the stage communiqué: finding out the formula explains the inscrutable, and fills the spectator with the pleasure of cognition; it also means aesthetic pleasure.[2] The work justifies its existence in the selected substance and form given, which is the feature of all works of art. This is why a work that rests on commonplace or the obvious cannot evoke cognitive tension, and, consequently, cannot lead to the pleasure of experience and cognition, and cannot justify its existence in the substance and form given. In the latter case, the same meaning may be achieved by the employment of other substances and other modelling. In the case of a work of art, however, none of the components may be substituted. If it is, the work is either damaged or loses its original meaning and aesthetics. The substitution of a good actor by a mediocre one, or a bad disposition by a good one, does not only influence the acting, but may destroy the whole work.[3]

We, the spectators, appreciate and admire a director's talent and imagination in devising new vehicles and configurations for the signifiers created in his/her performance. This, in a way, often becomes a criterion of the show's originality and artistry. As indicated above, the theatre as an artistic medium is always, at least partly, oriented towards itself. Theatre wants to draw our attention to its internal beauty, to the rules and materials with which it is constructed, and which justify its appearance in this particular shape and not in any other. As is always the case with artistic texts, theatre, as an exhibitionist and a narcissist combined, uncovers its anatomy, and allows us to contemplate its internal substance and composition, always unique, sometimes strikingly strange and surprising, but capable of conveying an abundance of information.

Elements reacting

If a chemical reaction was indeed staged in theatre (and there is no reason why it could not be), we could have, for instance, a scene presented in which two units of hydrogen merged with one unit of oxygen (2H+O), resulting in water (H_2O), played by three actors (or groups of actors/dancers/ puppets) respectively. Naturally, water would not have to materialize on the stage as a liquid substance. Quite opposite to chemistry and its laws, which make it impossible for us to see the particular elements in the compound (water), in theatre that is exactly the case: we can only imagine the result of the reaction (but not see the "real" thing, unless we fall under the illusion), but we can see its particular components taking part in the reaction. So, in the given example, on the stage we can only imagine the fictional water, signalled by, say, the gestures of actors impersonating hydrogen and oxygen, implying that the reaction is completed, accompanied by the change of sounds (babble of water), music and lights, but we can always see water's constituent ingredients, hydrogen and oxygen, which would not simply dissolve into a new compound as would have been the case in a true chemical reaction, but remain in their former material shape. This is a significant difference: in empirical reality both hydrogen and oxygen are invisible to us, whereas the effect of their "union", water, is; in the given example in theatre the opposite is true: the ingredients are and remain visible, and the effect of their reaction is not. And yet we know it is there! And it is worth noting that there is no iconic resemblance between the human bodies and the chemical gases that they denote: it is other signals emanating from the stage that make it possible for the spectator to read a human body as hydrogen. For instance, the fictional figures on the stage might describe it (through the utterances of the actors, of course) as if it were really there. So, what counts are the substances participating in the reaction and their relationship to the implicated outcome.

Also, what is needed is somebody's (who inhabits the fictional world) description of what the outcome of the stage reaction is. It does not have to be a verbal description: words may be accompanied by bodily movements, gestures, facial expression, etc. Please note that if the same reaction were repeated on the stage in the same performance that we would not need all the participating elements to appear again: a single, though conspicuous, element would suffice to stand for the whole; it could be the tune, light or some other component. This would provide the metonymy of the whole reaction. Also, in the world outside theatre, hydrogen will not combine with oxygen without a catalyst. In theatre, the chemical catalyst is replaced by ostension. It is the movements of the bodies, their changing spatial contiguity (blocking), and the accompanying sounds and words that make us realize, in this particular example, that the reaction is taking place and what the outcome is. The fictional figures clarify what otherwise would remain indeterminate. Even if real water appeared on the stage, it would remain an element of the set, a material signifier of the real water in the created world. This is quite contrary to the figures' perception, which usually see the final effect of "reactions" only, and do not see the "chemical" components. They do not see a semiotic world, but a real one. The result of the chemical reactions on the stage, prompted by the utterances and behaviour of the figures (signalled by the actors, of course), may develop and crystallize in the minds of the spectators only. It becomes a mental construct, a crystal if you will, which may be and is constantly confronted with the evolving stage text (and with our cognition of the world).

Of course, the meaning of the scene may be much more complex than just a chemical reaction, but in order to achieve deeper semantic significance, new relationships have to be signalled with the phenomenal world of the stage or the contextual world outside. A lot relies on the competence of the spectator, who has to be able to read various codes employed here. One needs to recognize only the rules of (theatre) chemistry in order to understand the scene presenting the reaction between hydrogen and oxygen yielding water, described above. But one needs to know the economical and political context of the performance in order to add an implied topical or metaphoric (allegorical) meaning to the scene and read it as a commentary on the need for, say, the political union of two parties or countries, which will result in the new spring of life-generating water (this would be a Peircean symbol). Even a royal wedding may be metaphorically expressed as a union of two elements, generating water, which is the source of all life.[4] Without the knowledge of the political and cultural context everything will be obscure, and the spectator will only be able to read the substance of the compound stage signifiers, without being able to see relationships between them or to grasp the referent. "There were fish, but no water", complained one of the naive spectators of an early seventeenth-century masque, presented at court, in which illusionary scenery, being the novelty then, showing an underwater kingdom, was used.

What is also fascinating in theatre is the fact that – again contrary to the laws of chemistry – the described reaction, and any other reaction, may be presented and contrived with different material ingredients, and through different stage actions or formulas (mime show, musical, dance, opera, puppet show, burlesque and the like), with the final referential effect being actually the same (i.e. water), although the formulaic, substantial and aesthetic side of the reaction would inevitably vary from one production to another. It would vary because of the different relationship between the signifier and the signified. Indeed, the major striking feature of theatre chemistry is the ability of different substances to react under different formulas in such a way that the result of denoted, primary or referential, meaning is similar, though never the same. At least it is similar in the sense of an object or the figure or the story or plot that unfolds on the stage. But the ultimate meaning in theatre is not the created fiction in itself, but also the relationship between that fiction and the substance of the signifier. Since the latter changes, so does the meaning and the aesthetics. The rule applies to the opposite situation too, when a single material component may under certain conditions, denote different objects, figures or phenomena. In one production, one actor can play a great number of strikingly different roles. This inevitably draws the spectators' attention to the unique selection and combination of substances and human bodies involved and to the formula employed, which results in the production of, say, *Ubu Roi* or *Twelfth Night*. As I have already pointed out, there is no table of elements applicable to the theatre, there are no discrete elements which are at everyone's disposal (such as phonemes or words in language), but there is an unbelievable multitude of possible materials with which theatre signifiers may be created, and their choice, shape and composition will enter into all sorts of relationships with the material substances of other signs and also with the world outside theatre. Let me repeat, then, theatre is a substance-specific medium.

Is there a dog on this stage?

In an attempt to show more precisely that different substances used to denote seemingly the same object or figure generate different meanings, I propose to discuss at some length a scene from William Shakespeare's play.[5] In a well-known scene in *Two Gentlemen of Verona* (act 2, scene 3), the clown Launce enters with his dog (Crab), which he reproaches for remaining indifferent to the misfortunes befalling its master:

> I think Crab, my dog, be the sourest-natured dog that lives: my mother weeping, my father wailing, my sister crying, our maid howling, our cat wringing her hands, and all our house in great perplexity; yet did not this cruel-hearted cur shed one tear. He is a stone, a very pebble-stone, and has no more pity in him than a dog. A Jew would have wept to have seen our parting. Why, my grandam, having no eyes, look you, wept herself blind at my parting ...

The above shows us how a soliloquy gradually evolves into a monologue addressed directly to the dog. This is explained by the fact that inner thoughts eventually take the form of a spoken reprimand. In what follows Launce says the following:

> Nay, I'll show you the manner of it. This shoe is my father. No, this left shoe is my father. No, no, this left shoe is my mother. Nay, this cannot be so, neither. Yes, it is so, it is so, it hath the worser sole. This shoe with the hole in it is my mother, and this my father. A vengeance on 't, there 'tis. Now, sir, this staff is my sister, for, look you, she is as white as a lily and as small as a wand. This hat is Nan our maid. I am the dog. No, the dog is himself, and I am the dog. O, the dog is me, and I am myself. Ay, so, so. Now come I to my father. "Father, your blessing." Now should not the shoe speak a word for weeping. Now should I kiss my father. Well, he weeps on. Now come I to my mother. O, that she could speak now, like a moved woman. Well, I kiss her. Why there 'tis. Here's my mother's breath up and down. Now come I to my sister. Mark the moan she makes. – Now the dog all this while sheds not a tear nor speaks a word. But see how I lay the dust with my tears.

Now, it is obvious that this is in fact the description of casting and directing a theatrical mini-performance, together with its inner show, more precisely – a puppet performance (entitled "My moving farewells with my family"), where the figures are created by shoes, a hat and a staff, and the only person (i.e., the human being) in this company plays … a dog. Let us also note that the only spectator of this scene – being in essence a play-within-a-play – is … a dog.[6] Shakespeare shows us with great humour that on the theatrical stage anything can become anything, provided the spectator is told as to who's who and what's what, that is when various objects and bodies are given sign-significance. We do not have to see Hamlet's ghost on the stage – it suffices that various stage figures see him. But Launce's performance is certainly not "a one-man show" as Kurt Schlueter would have it,[7] for apart from Launce, the dog and various objects take an active part in the performance, staged *ad hoc*, and they stand for various figures which appear in the microplay within. We also find here a monologue that is a description of the scenic action, of the behaviour of the figures, their characterization, and even various pieces of action (kissing the family) and a fragment of dialogue. We also have a division of the theatre space (and time) into two: the stage and the auditorium, and within the former there is created a world divided by time and space (this is why at least one spectator is needed – in this case it is Crab). And that is what we acknowledge as the fundamental feature of theatre. Thus, we have three time levels presented to us: the play within is level 1, and takes place in some indefinite past, when Launce was saying good-bye to his family; level 2 is the time of the performance staged by Launce in the fictional temporal reality of the macroplay; and level 3 is the time of the actual performance, that is, it is the time shared by actors (not figures), the dog (if live) and the spectators.

What is revealed here is the hiatus between what is verbally described to us by Launce, and what may be seen on the stage, between the substance and the denoted meaning of the stage signifiers (shoes, the hat and the staff become stage figures and the dog denotes a human being, and a human being denotes a dog); in other words we see stage synesthaesia in action. It is also worth noting that there is reason for the staging of this mini-play within: from Launce's point of view, the appellative (persuasive) function of the performance with the pair of shoes in the leading role is the most important here. The show is to persuade the spectator (that is, the dog) that it should change its indifferent attitude to that which concerns its master (similar is the intention of a number of full-length performances). In order to influence the scenic spectator (that is, the dog) more effectively than by a simple verbal reprimand, Launce *ad hoc* arranges a theatrical performance. The purpose is strictly illustrative and educational, which, again, is a feature of many plays. This comic scene demonstrates what is the whole paradigm of theatrical illusion: invention, direction, casting, acting, encoding, deciphering (or rather its impossibility because of the incompetence of the sole spectator) and convention. There is also faith, at least on Launce's side (not confirmed by the reactions of his spectator), in the power of the operation of the theatre and its educational function. What Shakespeare shows us here is his deep understanding of the essential features of theatre, but instead of providing us with a theorem, he enacts it on stage.

In his influential *Great Reckonings in Little Rooms*, Bert O. States uses the scene under discussion as an example to illustrate his theoretical argument. The passage in question is worth quoting extensively, for it touches upon some of the issues that are essential to my book.

> Anything the dog does – ignoring Launce, yawning, wagging its tail, forgetting its "lines" – becomes hilarious or cute because it is doglike ... We have an intersection of two independent and self-contained phenomenal chains – natural animal behaviour and culturally programmed human behaviour. The "flash" at the intersection, equivalent to the punch of a joke, comes in our attributing human qualities to a dog (a wagging tail is a signal that the dog has understood; a yawn is a signal that it is bored); but beneath this is our conscious awareness that the dog is a real dog reacting to what, for it, is simply another event in its dog's life. So we have an instance of Bergson's comic formula in reverse: the living encrusts itself on the mechanical – mechanical here meaning the prefabricated world of the play. In short, we have a real dog on an artificial street.[8]

However, I think it is worth pointing out that everything, whether knowingly or not, that finds itself on the stage, ceases to be itself and acquires sign significance or sign function. There is no such thing as the "real dog" in itself on the stage; equally, there is no such thing as an "artificial street" on the stage: everything is real in the sense of its material substance. A scenographic street, which phenomenologically is "real", no matter what its iconic similarity to a "real" street is, becomes a sign of a fictional street which is inaccessible to our senses. Everything is „artificial" in the sense that it is conventional, i.e. conveys referential meaning that is entirely dependent on the mutual agreement between the performers and the spectators. So, there is no difference between the "realness" of an object or a dog or an actor's body. The difference lies in how they are perceived by fictional figures and whether or not these items, if live, have the capacity of signalling their consciousness of their own "I", of time and space (i.e. their *deixis*). The dog in this particular case is not funny because it is a "real dog on an artificial street", but because it is expected to reveal that capacity of signalling attributes of human consciousness, and is treated by Launce, who wants to establish chemistry between him and the dog, as his generic equal (and, naturally, fails to fulfil these expectations, quite unlike us, the real spectators). Launce knows perfectly well that in order to create a theatrical situation, at least one spectator is needed. Since there are no humans around, he assigns the role to the dog. But what counts is that Launce-the-figure sees the dog; that leaves no doubt that the dog-as-performer is not only a "real dog", but also a sign of a dog living in sixteenth-century Verona.

The dog, indeed, is a complex sign and it is cast in several simultaneous roles: it is a dog in time level (1), that is, it is itself from the past in the play within, it is a dog and a spectator in the fictional reality of *Two Gentlemen of Verona* – time level (2), and it is also a real dog performer in the actual staging of the play (time level 3). And it is also, Launce tells us, a sign of him, at least momentarily. Similarly, Launce is the visible director and "animator" of the entire scene, and simultaneously he impersonates the Launce from his own past and is also a sign of the dog, enacted in the seeming here and now of the implied audience.[9] The only difference between the two performers is that unlike his master, the dog cannot reveal or signal its consciousness of the past and of a different space. It cannot be an actor, consciously and knowingly creating an assigned role. It cannot intentionally "impersonate" a fictional dog. It cannot be the spectator in theatrical sense, for it is incapable of recognizing and blending the input spaces, it cannot read the performance text. It is only the human mind that has that temporal and spatial consciousness along with the verbal and gestural ability to express it. This is why performers may in fact be inanimate objects (as in puppet theatre) or animate but non-human (animals), but as "actors" they have to be equipped with that ability to express the consciousness of time and space. As noted earlier in this book, there has to be a juxtaposition and a clash between what we, the spectators, see and hear on the stage, and how the figures see and hear and describe the same objects, bodies, sounds, words and so on (all the material substance of the performance). Without it no theatre is possible. Of course, what we have here is yet another model of perceiving reality – that of the dog; but what is noticeable is that without the mental and histrionic ability to pretend, we may assume that the dog perceives the stage world as part of its non-fictional realm. So, this creates yet another level of theatrical conventionality: the dog is given sign-significance, by which it becomes the sign of a fictional dog, living in the same time and space as Launce, but it does not recognize that reality as fictional.[10] Hence the impossibility of its comprehension of the performance as a semiotic construct and inability to convey to the audience signals of a model of perceiving reality different from the dog in the street's. Similarly, a dog may be a witness of an event, but cannot be a spectator. For this reason the dog in the play under discussion may be seen as an actor only in some limited sense (because it fails for reasons I shall describe below) provided it is equipped with that consciousness by its animator, that is, Launce, whose verbal and gestural utterances inform us that the action of his mini-play takes place "earlier" and at a different space. Without him, the dog could not signal to us that it is in fact enacting itself from its own past or that it is a sign of Launce in the past (stage acting cannot be carried out unknowingly). This is also the scene's humour: the dog sits down, watches and listens as we do in the theatre, but it does not comprehend the dual ontology of everything which is presented to it. It is in the extreme a naive spectator, or, rather, reminds us of someone who has by chance found him or herself in the field of a play or game and does not recognize the rules that make the space different. It cannot see what Launce wants it to see: it will never see crying shoes, will not smell a sign of Launce's sister in the staff, and the smell of the shoes will not remind it of Launce's mother. This in fact is the source of humour in the play: we find ourselves in a position to be capable of accepting the illusion created by Launce's little theatre, but the dog's incapacity to accept or signal conventions of theatre reminds us of the illusionary nature of the art and unveils the convention. Breaking the illusion, violating the rules of the theatre, whether intended or not, may in certain circumstances be metatheatrical and funny.

All of that brings us to yet another problem, because it becomes apparent that a live dog (any animal) on the stage is basically untheatrical. Why is that, one may ask? Well, precisely because it is alive. It is a biological clock. The very fact of its being alive means that it exists in the spectators' present, and cannot appear, without pretending, as living in the past. But it is only a human actor that has that ability, and can pretend to appear in the past as if in the present, that is, in what I have labelled the transferred present. With the "real" dog present on the stage as a vehicle of a sign of Crab, there is a conspicuous discrepancy between what Launce is saying and what he is expecting the dog-as-an-actor to signal, and what the dog-as-an-actor-to-be really does on the stage (in all sorts of variants that have occurred in the stage history of the play). This discrepancy, resulting in the dog's failure to become an actor (an inability to impersonate its own fictional species), is an additional source of humour in the play (lost in the case of a puppet or "invisible" immaterial dog, which by definition cannot act themselves). The discrepancy is partly caused by the fact that the dog's "natural" reactions to the world around it, its behaviour which is in fact bereft of any artificiality or acting, place it in time level 3, congruous with the temporal reality of the audience and the actor. So it can see the actor, who is also in time level 3, but cannot see Launce, who is in time level 2. Similarly, Launce perceives Crab, but cannot see the "real dog". The live dog as a phenomenological being belongs to the spectators' reality; however, since it phenomenologically shares spatial and temporal, hence indexical, relations with a fictional figure, unknowingly (the "candid camera effect") it becomes a sign of a fictional dog living in the sixteenth-century Verona. In other words it is the context that allocates sign-significance to the dog, and the latter unknowingly ceases "to be itself".[11] So there is a conspicuous clash between the live dog-as-performer and its unknowing function as a sign of something else, resulting from the mechanism or the conventions and laws of theatre that govern that context. One of the basic rules of the theatre is that both the performer and the spectator take part in an act of communication on a voluntary basis. Thus, any unknowing performer is bound to be funny or ridiculous or both, because what we see is a being entangled in the mechanics of an evolving spectacle that remains incomprehensible to it (this feature is often exploited by theatre companies that attempt to force a spectator to become part of the show). Whatever the dog does may appear as funny. Instead of the clash of two models of perceiving reality, we see a performer incapable of signalling anything different from what we see on the stage. The chemistry of the theatre cannot take place, for it lacks one of the retorts taking part in the reactions. If it were otherwise, in the scene under discussion we would have a typical theatre-within-the-theatre convention employed. As it is, the realness of the dog breaks the illusion of fictionality.

Moreover, the dog as a real thing cannot read the mini-performance as a medium, but only as part of its real world, in its here and now (time level 3, as we perceive it). It cannot be an active participant in an act of communication staged by Launce and by Shakespeare (and the director in theatre performance). So the humour of the scene is partly dependent on the real presence of a "naive spectator" (i.e. the dog) who does not reveal any ability to read the stage text as a play. We can see that the dog does not really understand what is going on within the performance (of course it has its own way of perceiving the world, but one which is beyond our comprehension). And the fact that it cannot read what we read and Launce reads is the source of humour: the dog is funny because in spite of instructions it cannot grasp Launce's intentions of creating a fictional reality, so it becomes a sign of naive spectator who is unable to differentiate between the stage fiction and the world around him. An additional clash between different models of perceiving reality is created, namely, that between a conscientious spectator and a naive one. What is funny is the predictable futility of the effort to create a real spectator out of a dog, proven by the latter's reactions (or lack of signalled understanding). The dog may be given a role, but cannot be an actor by itself. Its acting may only be attributed by a human agent. Launce does not seem to understand that and his high expectations are bound to fail, which is yet another source of the comic in this scene (not to mention his expectation that the dog will cry).

Let us now consider other possible ways of staging the scene, without a live dog. If for instance the dog were performed by a puppet, something would be lost, especially if Launce followed the rules of the art by adjusting the movements ("acting") of the puppet to his own words. In that case we would have a "true" mini puppet show (with other objects being the remaining puppets). Consequently, the source of humour would change. So, if Launce addressed a puppet as if it were a live dog, we would instantaneously observe that inadequate material, obvious to us, remains adequate to the clown. We see what he ought to be seeing. All the reactions and behaviour of the dog are assigned to the puppet dog verbally, but in this case are not negated by the fact that it lives in our own reality, in time level 3, because it is not a live being. It is totally fictional, and can convey any meaning assigned to it. So, something else is funny: it is the clash of two contradictory models of perceiving reality: Launce's, who sees the real dog with whom he converses, and ours, the spectators', who see the puppet which is not live, a material substance of a sign of a dog. Another variant of the scene would be if Launce recognized the puppet in the dog. In that case, he would appear to be playing games with himself. Moreover, the puppet on the stage may reveal iconic similarity to another dog known to the spectators from the media, films or even politics (i.e. the President's dog), and through this create additional meanings.

Yet another possibility is that Crab is played by an actor as in the 1952 Bristol Old Vic production directed by Denis Carey. In this case we have a human being who is a sign of a real dog (time level 3) who is a sign of a fictional dog (time level 2) who, depending on Launce's casting, is in turn a sign of a spectator or itself in its own past (time level 1). In this case there is no discrepancy between the behaviour of the actor and the behaviour of the dog that he/she enacts. We also know that the actor's presence on the stage is voluntary, and that he/she has the ability to signal the sense of time and space, but whether or not he/she equips the dog with that ability depends on the director's decisions. So, generic or biological inability is replaced by conscious choice. The actor can follow the intended behaviour of the dog suggested by Launce's utterances, or do something to the contrary (he can even cry during or after the show within). Even if the latter happens, we, the spectators, recognize the presence of a human being impersonating the dog, and for this reason we realize that whatever happens results not from the lack of understanding, but from conscious decisions undertaken by the actor. So part of the humour is lost, although it is supplemented to some degree by the fact that a human being playing a dog is funny in itself (one of the options in Launce's casting is that he himself plays the dog, while the latter plays the spectator[12]). It is funny because a human body – even without the raising of the leg – is entangled in the alien mechanics of animal movements (in a Jacobean play, *The Witch of Edmonton*, one of the two leading speaking parts is played by a dog).

This also explains why we could actually have the same scene enacted without the physical presence of a dog: Launce could with ease signal to us (verbally and through acting, with or without a leash in his hand) that the fictional dog is present on the stage "now" and/or in his past and at a different space. A fictional figure can be created even if its signifier is not physically accessible to our senses (another actor can signal its presence in the fictional world). In point of fact, Launce begins his speech by introducing the dog ("I think Crab, my dog ... "), so the presence of the dog, invisible to spectators, could be marked by his acting, words, gestures, gaze and so forth.[13] What would be lost, however, is the real presence of the live dog as a sign of a figure of a dog in the fictional level 2 (the macroplay) – the one that pretends to be simultaneous with the time of the implied audience as if temporal levels 2 and 3 merged.[14] What would also be lost is the clash of two different models of perceiving reality presented to us on the stage: the "real" Launce's and the "real" dog's. That is to say, no immaterial and, to the spectators, invisible presence can mark or signal the slightest sign of consciousness or lack of consciousness (this has to be signalled, if need be, by the accompanying conscious *ego*, visible and audible to spectators). Launce as an actor and as a human being has no problems with "reading" the play he directs and enacts, nor with signalling to us, the spectators, that the time and space of what we see is ontologically split or doubled (the one is semiotic, the other implied). His created world remains sensorily inaccessible to us, and what we see and hear are only conspicuously inadequate signs of the fictional reality (as in Plato's cave).

These examples have shown that the signifier, the material vehicle of the sign, has a considerable impact on the semantics of theatre, and every change of the used substance leads to the change of the semantic structure of a given production, its aesthetics and humour. This explains why Launce is at odds with himself when trying to make final decisions as to the casting of his play within ("I am the dog. No, the dog is himself, and I am the dog. O, the dog is me, and I am myself. Ay, so, so."). The consequence of what Shakespeare is showing us and what I have been trying to describe is the ability in theatre of anything to become a sign of anything, provided there is at least one *ego* that reads the material substance of signs differently from what they are and mean in extra-theatrical reality (which ontologically is uniform). The point is that the substance of theatrical signifiers does not have to be similar to its designated meanings, and Shakespeare shows us clearly that even "aeiry nothing" is an adequate substance. However, quite contrary to what Bert States writes about this scene, the dog is not a "dog-in-itself"; it is a dog not only because we see it and interpret it as a dog, but because a stage fictional figure, Launce, reads the dog as a dog (equally he could read it as his human friend, and Shakespeare shows us that in theatre even that is possible). So, even if the ontological or phenomenological reality of the dog is obvious, it does not mean that it cannot become a sign of something or someone else. In fact it remains, throughout this scene, an unknowing sign of a fictional dog inhabiting the world of the play: there are two dogs here, the real one and the fictional. It has a split nature, and it is both real and semiotic, like everything else on stage. This duality is the essential factor in theatre, and I have defined it as theatricality. The condition is, however, that there is a signalled clash between at least two models of perceiving reality (between the spectators' and the stage figures'). As I have observed – without it no theatre is possible. And Shakespeare shows us that condition in varying versions, unveiling the mechanism of theatre. In the example given, the fictional dog is not denoted entirely by the real dog, which in this sense fails as an actor, but by the actor playing Launce.

We have seen that different objects have the ability to denote the same thing: in this particular case a sixteenth-century dog, Crab. Even its material absence turned out not to be an obstacle that would make the signification impossible. As discussed above, whenever one denoted object or being is denoted by different signifiers of different material composition, the attention of the spectator is drawn towards the nature of theatre signs rather, than to the elements of the fictional realm (which is the case when the same scenic object denotes different objects or beings beyond the fifth wall). In other words, it foregrounds the conventionality of theatre, and by doing that undermines any illusion that may have arisen in the process of perception. It is highly unillusionistic. Consequently, it may become an additional source of humour: we are laughing, because we show that we cannot be fooled in believing the unreal.

Theatrical osmosis

Metaphorically it could be said that the relation between the two fundamental spaces of the theatre, the space of the stage and the auditorium, can be compared to the phenomenon of *osmosis*: separated by an invisible membrane, they pass on the signals of the stage communiqué from one sphere to the other, from one retort to another. Osmosis also separates and selects elements of what is seen and heard on the stage and draws our attention to what at a given moment is taking part in the intended chemical reaction. In some sense this is reminiscent of the mythological gates of horn, through which only dreams can pass (as indicated in Thomas Kyd's *Spanish Tragedy*),[15] or of the role played by a conductor of an orchestra. Theatrical osmosis enables the selection and combination of particular components leading to the formation of a compound signifier, points to their reactions, establishes relationships between them individually and between the whole sign and whatever precedes and follows it. It takes an active role in the transmission of data from the stage to the mind of the spectator, initiating and directing the latter's perception processes, which includes memory. On the surface it seems that it is the spectator who freely selects and combines various components of the performance text, visible and audible at the same time within the space of the stage, reacting with one another. But in fact it is the director and other creators of the show who prompt the spectator which of the sundry possibilities to select and combine in a given time. In this way they control the perceptive and cognitive process, and make sure that the spectator concentrates on what they think is proper. If we select wrong components, the outcome will inevitably be different from that intended. Once the stage signals find themselves in the mental retort, they take part in further reactions, constructing the blend, which is then related back to the stage as one of the input spaces, which, in turn, is confronted with the changing space of the stage, resulting in a new blend in a continued reception process. Thus it is vital for the creators of a theatre production that the spectators "put together" the intended components, and do not mistakenly blend them with others, which are made redundant at a given moment.

The metaphoric membrane of the process of osmosis is however very delicate and easy to tear, for instance by bad acting. It has to be remembered, however, that the latter may have other causes, such as incompetence of the spectator, his/her psychic disposition, presuppositions, etc. Thus it is not only that which is happening on the stage, which can influence the process of osmosis. For the porosity of the membrane, its "throughput", is dependent not only on the director, but also on the performers, on the spectator and the context of the performance. So, on the one hand, it is being bombarded by various loads of signals and information from the stage, which – in accordance with the will of its creators – greatly desire to be foregrounded and thus acquire sign values by squeezing through the pores of the magic membrane (in various preordained configurations, marked by the "filters" used), and prompt the spectator which of the input signals emanating from the stage are to be blended together, i.e. which of the components enter, at a given moment, into reactions in the stage retort; on the other hand, whether or not this succeeds in conveying the message is dependent on the spectators.

The process of osmosis has to be under control and to bring order, the filters of ostension are used,[16] which allow us to order everything that passes through on two principle axes, the axis of combination or contiguity (syntagmatic) and the axis of selection (paradigmatic). As indicated above, theatre is a spatial medium, so, naturally, the spatial composition of individual stage pictures has to be taken into account. Of the multitude of possible combinations and relations between the multitude of the components of the stage, the director always selects and draws our attention to those which at a given moment enter into intended, hence programmed, mutual relations or reactions, whereas other components are at rest, so to say, waiting for their cues. To achieve this, the director may use light, colour, sound, the actors' use of the language, gestures and gaze, and so on, all of which activate some of the components by bringing them together, neutralizing the others. The importance of light cannot be overestimated, although the spectators might not be conscious of the ways in which light influences their vision and mental states in the process of perception. Illumination is only one of the several functions of stage light, the others being sculpturing the actor's body and its relationship to the scene, selectivity – which is one of the most basic ostensive techniques, and fluidity of changes.[17] In the course of the twentieth century light becomes integrated with the dramatic action, with the psychic states of fictional figures, and thus became an inseparable component of stage reactions.

Ostension has the ability to pass the systemic boundaries, and brings together elements that in normal circumstances do not form any relationships. Moreover, it has the ability to merge elements that are divided by space and time, for basically what ostension does is that it shows us the constructive elements of the imaginary structure or blend that we, the spectators, are expected to create in our minds. Thus, directors and actors are at pains to keep control over the spectator's perception, in respect of which of the many stage components he/she concentrates on. This is important, because the spectator is expected to notice the components of a particular scenic amalgam, grasp the scenic reactions between various, but specific, substances and, consequently, grasp the intended meaning of a scene or utterance. It is ostension that establishes the temporal and spatial contiguity or indexical relationships between various components taking part in stage reactions. Thanks to ostension, words become integral parts of heterogeneous signifiers, which constitute units of a higher order, different for every production, even if the same dramatic text is used. In this way the conglomerate substances of the signifiers have a great impact on the semantics of signs, which may be seen as a speaking modelling clay. Words enter into these units with their semantics from outside theatre, and, when incorporated, become semiotically and semantically transformed into stage speech (see Chapter 6).

Ostension is similar to the way a conductor conducts an orchestra, selecting the instruments that play simultaneously and those that are, for the time being, silenced. The theatre director, however, instead of a baton uses various "filters", modes of ostension. This also reminds one of chemical reactions, in which out of simple and unrelated elements, such as gas and metal, compounds of different shape, structure and composition are created, revealing different physical features. Moreover, they reveal the ability to convey meaning. Ostension may also be seen as a catalyst which allows given components to react with one another, which, in other conditions, i.e. without the catalyst, would not have reacted, because their "normal" relationship is not spatial (based on contiguity), nor is it based on similarity (iconicity) or sequential relations (indexical cause and effect). The same function is revealed through paradigmatic relations, by which the spectator links and blends signals from different scenes, involving different actors, words, objects, light, etc. In the blending process, the real spatial and temporal distance often becomes irrelevant. Similarly, the intended composition of a stage picture, in which a number of components appear simultaneously in relationships that can be identified by the spectators as not accidental, is a form of ostension. The modelling brings objects, sounds, light, and human bodies together, by which they gain the ability to convey meanings that otherwise could not be conveyed by individual components.

What this means is that quite contrary to science, the stage reactions may take place even if the individual components are separated by time and space. This is made possible through the use of two retorts, the phenomenal on the stage, and the mental one in the mind of the spectator. The reactions in the former are transitory, set in streams of fictional time, and evolve from one into another in a sequential order; the reactions in the latter are not transitory, but, based on memory, accumulate to build a structure of syntagmatic and, more importantly, paradigmatic relationships. This results from the fact, mentioned above, that the performance does not have its "past" when unfolding in front of our eyes and treated as an artistic work in its wholeness: it is a finished whole that has to show all its constituent elements in order to reveal its structure and convey the message. In the theatre, it is not the present that determines the future, because the latter is totally pre-ordained. It is the future, which is given *a priori*, that finds its phenomenal realization in the present of the performance. Thus, we recognize the existence of a mental retort, in which various input components are consecutively brought together within the space of memory to react and blend with one another.[18]

All of this implies that the structure growing within the spectator's mind, being the result of the perceptive process, is not a mirror image of what is occurring on the stage. It is also not a mirror image of what is implied in the fictional realm. The spectator is expected to notice the structure of the whole, i.e. a network of relationships between all components. Thus, the ultimate meaning arises in the mind of the spectator, in very complex reactions taking place in the mental retort. The spectator's memory is always taken into account by those responsible for a given production. To save time, they count on the spectator's ability to notice equivalences and relationships that are not at all obvious and are not signalled simultaneously by the stage text. Osmosis prompts the spectator what to store in memory, and what may be neglected or erased. It prompts the spectator which elements of the signalling matter are to be brought together. Moreover, once a given reaction is shown to the spectators, in later scenes it suffices to bring only one of the ingredients on to the stage, and it becomes its synecdoche, a material fraction of the whole sign (the process may be called synechdocal multiplication), a sign of a sign of a sign, if you will. Thus complex reactions (and their meaning) may be evoked by a simple ostensive appearance of a single object, word or sound. If the Weird Sisters in *Macbeth* first appeared on the stage in a strange light, accompanied by a distinctive tune, then in the following scenes they could appear without the physical presence of the actresses, signalled by the light and sound only (being now their metonymy). This could make their presence at Macbeth's castle more frequent or even permanent, if such were the artistic vision of the director.

Moreover, all the simultaneous signals emanating from the stage are filtered in the mental process of osmosis, by which some fall out, while others enter the space of memory. Thus they become active components of further mental reactions in the individual retort of the spectator, where the resultant structure of the whole work is being constructed. They also become components of mental blends, which are constantly confronted with new signals and pieces of information emanating from the stage, in a process reminding that of a cognitive kaleidoscope. Wrong choices made during the process of osmosis, wrong selection of components entering the space of memory, results in fallacious blends and are responsible for interpretative errors. Bad directorship may result in the spectator's inability to combine the desired components together, by which they will not react with one another. In either case, the mental space of memory is a *sine qua non* for the rise of a coherent structure of the whole work. This is theatre chemistry or communication in action and allows even the simplest signs (in material sense) to become rich in meaning, hence to convey and store more information.

Stage *amphoterism*

As indicated above, different materials, objects and bodies, engaged in different relationships may denote seemingly the same figure, object or state of being in the fictional realm. But by now we know that the meaning of the signified and the aesthetics of the whole will always depend on the material used and the way it is modelled as the signifier. Moreover, the same object, a prop, may denote different things in the fictional realm. Again, the selection, modelling and spatial arrangement of simple substances influences the semantics and aesthetics of the performance. This feature of theatre signs is similar to the phenomenon of *amphoterism* (from Greek *amphoteros*, meaning "double") in chemistry, whereby certain compounds may function as either acids or as bases, depending on the reactions they are engaged in. There is no reason why we could not employ this term to label the peculiar feature and ability of singular theatre signifiers, substantially unchanged, to denote different things. They would then be labelled *amphoterous*. Thus, one object (or the same material different objects are made of) may become a sign of two or more different things; in a similar way, one actor may have two or more roles in one play. This also means that more than one object may become a sign of just one thing, and two or more actors may enact the same role, depending on the reactions they are engaged in. In the 1920 Nikolai Evreinov's *The Storming of the Winter Palace* in Leningrad an outdoor theatrical reconstruction of the symbolic event of the Bolshevik revolution, which took place on the site, with the participation of 8,000 performers and around 100,000 spectators, there were 35 actors playing Kerensky and 20 playing Lenin.

Thus, amphoterism also applies to the actor, who can change his/her sign value and orientation during the same performance. In 1906, a Dutch actor named Henri De Vries presented a sketch in which he played seven "characters".[19] In Roberto Bacci's *Hamlet* (2008), the cast of seven actors assumes changing roles: the same actress may be Gertrude, and in the following scene she will be playing Ophelia. Apart from the previously mentioned situations, when the actor becomes the sign of an object or animal, one can recall the famous staging of *King Lear* by Ingmar Bergman from the middle of the 1980s, performed in many European countries, in which the body of the actor could have become a figure, only to become transformed after a while into the sign of an object. Thus the signs creating almost the whole construction and scenery of the stage were created by the actors and their bodies. When Lear sat, the chair (seat) was the back of an actor kneeling on all fours; another actor was the table, and yet another – the wall. The same actors after a while became the signs of figures, then after another moment – after playing out the episode – signified something else again. In this way all the actors were permanently present on the stage, but the configurations of the figures and the objects, as well as the time and spaces, were changing. This extreme example shows that the material of a theatrical sign really has within itself a great potential for transformation, and theatrologists see in this the essence of its theatricality.[20] A similar thing occurred in the staging of *The Beggars' Opera* in the Swan Theatre in Stratford-upon-Avon a few years ago: there was a scene in which the bodies of the actors formed a speeding coach, with horses, coachman, carriage with passengers, and even – through the use of music, appropriate gestures and body movement – denoted speed, change of space and even the bumpy road (the same staging trick was applied earlier in the almost nine-hour-long *Nicholas Nickleby*, directed in 1980 by Trevor Nunn with the same RSC). In passing it could be said that, thanks to amphoterism, sign transformations depend also on the change of time series: plot signs become non-plot signs and vice versa; or active signs (i.e. those foregounded by ostensive means) become inactive and vice versa. Crossing from one time sequence to another is a signifying signal within the stage text. In the quoted example of Bergman's production of *Hamlet*, the body of the actor – the substance of the sign of a figure, which forms with others the action of the plot, which is moving in its own time – undergoes reification, a change into an object, which is an element of the scenographic construction of the space and as such does not succumb to a visible change in time, or at least does not signal such a change. Something different happens again in Caryl Churchill's *Cloud Nine*, where we have an unconventional cast: in the first part of the play a girl, Betty, is played by a man, black Joshua by a white man, Edward by a woman, and a child – Victoria – by a puppet. In the second part all the figures are played by the "appropriate" actors, in accordance with tradition. These metamorphoses are not exclusively theatrical showiness, but they create meanings which are legible within the whole work.

As indicated, in theatre objects may become signs of human beings, and human beings may become signs of objects. A human body, carrying a bundle of sticks, along with a lamp and a dog, become a sign of the Moon in the play-within-the play in Shakespeare's *Midsummer Night's Dream*.[21] In another case, in scene one a wooden stick may mean a walking stick (being a sign of older age, in itself having symbolic connotations), in scene two a dagger, with which someone is murdered, and in the next one a sceptre, being also a sign of royalty (we now understand that the old person has murdered someone and has become a king); in scene four, it may mean a sword with which the king's opponent is beheaded; in the next scene it may become a sign of a wine glass (as in Robert Wilson's *Woyzeck*), when a toast of victory is raised with wooden sticks (the click of glasses is heard from offstage). In each case, the particular meaning is assigned to the same material object through words and acting (gestures, gaze, etc.), and through this the object draws our, the spectators', attention towards itself with additional force. Its seeming inadequacy is multiplied, hence becomes more conspicuous, and it absorbs its consecutive meanings.

However, through its fluid semiotic engagement, the same object also draws our attention to the denoted objects: it brings them together, so to say, creates equivalences on the paradigmatic axis. They have, after all, at least one thing in common: the substance of their signifier (as Brook's bricks, mentioned above). Through that linkage, the denoted objects, events or phenomena create new meanings, which may reinforce the action or message of the play, in which the sign of old age becomes a sign of cunning intrigue, and of surreptitiously obtained power which, consequently, is converted into tyranny, which, in turn, is the source of the tyrant's joy. Moreover, all of these meanings become inscribed on the simple object used, i.e. on the wooden stick, which gains the semantic richness not to be found in the world outside the stage. In this way, selected objects may reveal their amphoterism and accumulate meanings on various levels and employing various codes. If, in the example given, the play ends with someone breaking the stick or throwing it into a fireplace, the stick may gain further symbolic meaning. Equally symbolic would be a final scene, in which a new figure would appear at the tyrant's court, and that figure would make its entry walking in with the help of a walking stick. On yet another level, it may also reveal a topical relevance, alluding to a living tyrant. Moreover, objects may extend their lives beyond the play in which they first appeared, and emerge in another play, carrying the meaning obtained in the previous one[22] or alluding to a different text in which they became known (such as a film, a television commercial, a bill-board or a video-clip). What is close to the miraculous in the chemistry of the theatre is the unbelievable potential of its signs to absorb and convey information. Let me repeat, on the stage, as in an alchemist's dream, everything may become gold if it glitters; moreover, it may be gold even if it does not glitter.

Part II
The Chemical Laboratory

4
Sculpting the Space, or the Retorts at Play

Natural space vs. oriented space

The argument behind my initial assumptions is that spatial and temporal structures and relationships are the basic way in which individuals and whole societies both shape and comprehend the world around them. One should add that these structures and relationships may be of a dual nature: physical, that is, belonging to the empirical world, and mental, that is, belonging to imaginary, fictional or metaphysical realms. There can be no doubt that in everyday experience we perceive space thanks to the possibility of movement. Since it is people who move in space, and not the other way round, they become responsible for the appearance of a temporal dimension in their cognition of the world around them. Even if this sometimes happens against our will, it does not change the general principle: it is not the new space which comes to us but we who go to it. The space which is available to our senses remains static, stable and invariable, and, not being in motion, is *amorphous* and ahistorical. Natural space is a scalar, which means that it is directionless. Also, it is not grounded in some concrete time (I omit here the spaces created by man), since space is historical only when it has orientation or a privileged perspective; in other words, space gains a temporal dimension when it is modelled. Natural space has no orientation (it is not intentionally shaped for our reception) and at most people can define the place from which it is perceived, where the fundamental points of reference are our eyes and bodies (and our deixis). Let us notice therefore that in the case of natural space we can talk about the vector of looking, and not about the orientation of the space. For this reason, in order to orient in an amorphous space, humans define it according to conceptual patterns, such as the point of the compass, latitude and longitude or a map. In the complex perceptive processes, the natural space is transformed into an imaginary construct, a cognitive space, which becomes involved in acts of further blending, memory, imagination and so on.

The phenomenon of perceiving space is completely different in the theatre. It is not we who display activity, but the scenic space of the stage, which – in changing scenes – comes to us. The actors make their entrances and exits, the scenery changes, so does the light, and all within the space of the stage. And thus we do not experience or notice physical relocations in space. We do not even have such a possibility (with rare exceptions). It is not we who impose an orientation on it, but it appears to us as already oriented and modelled by someone; it is also *anisotropic* (varied, changeable in its dynamics). Thus in the face of our, the spectators', total stasis (with a tendency in today's theatre to move audiences around different spaces), we are presented, through the succession of scenes, with changing, therefore, dynamic spaces filled with various compositions and configurations of human bodies, objects and light, i.e. of compound signs. A garden scene is succeeded by a scene in the dining room of a manor-house (which we never see in its entirety), and then we move to a bedroom or a barn or a pond, or rather they "come" to us. In order for these spaces to become verisimilitudinous in their material appearance (if such is the intention of their creators, which is not necessarily the case), the interior of the box of the stage is built up in various ways, sometimes like a doll's house for huge children.[1] Furniture, often looking like the real thing, is placed together with walls and doors leading to other unseen rooms and numerous objects, both usable and decorative, from carpets to flowers in vases and pictures on the walls. An illusion can also be created on the stage of natural open spaces, stretching almost to "infinity". All this can resemble a three-dimensional painting, sometimes an abstract one, an artistic installation, or even a photograph. And, of course, natural space, in its manifold shape and appearance, may also be used for theatrical purposes. A number of productions in today's world use screens and monitors, film, video and computers that co-create and model the space of the stage.[2] The stage itself constitutes a material separate element (or, rather, an extraordinary "speech-apparatus") of whatever space is used for the performance, but at the same time during the performance it can become the signifier of some other place: the floor of a bourgeois salon, banks of a castle, a country road, the surface of the moon, the deck of a ship, a ceiling, or even "an upside-down stage".[3]

When we look at a natural (real) space, our mind wanders freely and what is more can move backwards or forwards in time. In the theatre this is impossible, since the oriented space of the stage is connected indivisibly with the preordained though varied stage time, imposed on us by the creators of the performance. We can, of course, in our mental processes move backwards and forwards within the limits of the fictional time of performance, through recollecting events and utterances that have taken place, and anticipating those that are the fictional future. Space in the theatre ceases to be a scalar. This concerns also the implied natural spaces, the whole fictional world. This testifies that this is *de facto* a projection of someone's view of reality, its model (with the character of a holistic *universum*), a materialized cognitive blend, which is presented to us to view visually and acoustically or to stimulate the imagination. Thus, the inhabited space of the stage may be seen as an exteriorization of its creators' minds and their imaginary "language", by which they want to convey their message or artistic experience. Since the space of the stage may be treated as a cognitive model exteriorized, it does not have to be mimetic; it may, and often does, resemble an imaginary composition, which does not have an extension outside the stage. This projection is intentionally shaped or modelled for our reception, is grounded in specific – although also fictional – time, actually a whole network of different times, and it is this which gives it orientation. What is more, we know that that which cannot be seen on the stage, namely the diegetic (implicated or described) space, in reality does not exist in a material sense, it is only an element of the fictional world. In this sense this is a physically discontinuous and limited space, not only because of the dimensions of the stage, but also because that which goes beyond the fifth wall is only implied, and we know (I ignore the naïve recipient) that this is quite simply fiction and can exist only as a mental construct in the mind of the spectator.[4] I propose to distinguish at least five theatre spaces: first, the space of the theatre as a building or, more generally, an architectural edifice; second, the space of the stage in its physical appearance; third, the fictional space that is denoted by signifiers, like the stage set, lighting, music, the actor's body and acting and, of course, the spoken word; fourth, the space of the auditorium, often neglected in scholarship; and fifth comes the immaterial mental space in spectators, where the previous spaces become blended, resulting in a cognitive blend.

Thus in the theatre the fictional space in its very essence does not correspond to the natural space, and the fragmentary (although in itself cohesive and continuous) world which it does embrace is modelled and governed by laws of physics which are different from those known from our experience. In spite of certain similarities, almost everything visible and audible on the stage foregrounds its difference from the empirical reality off-stage. Thanks to the orientation and modelling of the space, it ceases to be amorphous, although its relations with the points of the compass, latitude and longitude are completely without significance (they might be significant in rituals), unless they become, for some reason, relevant to the creators of a given performance. The points of its attachment (and reference) are elements of the material substance of the performance. That is why the stage sunrise can be in the (real) west, the Earth can be flat and no one pays any heed. From this point of attachment, in principle the focus of the viewing direction of the whole audience, comes the feeling (in spectators) of community of reception and the unity of space; in order to achieve this unity, walls and ceilings are not required – one material point, which will concentrate the attention of a group of people, will suffice. This has been demonstrated by detailed psychological research. I need not add that this point in theatres of all types is the stage (even in its basic form of a plane). That is why the creation – even in the simplest of forms – of a material sign of a body is such an important factor, and two perpendicular planes, of which one is the floor of the stage and the other is the back wall, are quite sufficient for this purpose, as theatre history shows us.

Thus, among other creations of the human mind and technical skills, theatre and drama may be seen as a reflection of a particular cognitive, spatial model of the universe, created in a given period. Furthermore, the theatre itself – as a building or any other kind of space in which a performance takes place – will usually emphasize its otherness and also lead us to a correct deciphering of what constitutes its interior and sometimes also of what we can see on the stage. This brings us to yet another problem: do theatres as buildings have meaning? If the answer is yes (and I am inclined to think so) then it follows that the external shape and the significant physical space of the auditorium may be the components constituting our understanding of the spectacle: we watch the play through the prism of the meaningful "text", or ideological/iconographic programme of the theatre building and interior. What follows is that even the same performance will generate different meanings when staged in different spaces. Stuart masques, for instance, gained specific meanings when staged in Inigo Jones's Banqueting Hall, where the ceiling, painted by Rubens, presented an iconographic image of Stuart ideology which the masques corroborated.[5] In the ideological and iconographic programme of the Whitehall Palace, its space was identified with the Temple of Wisdom, Peace and Honour. In George Chapman's *Memorable Masque* (1613), the Temple was in fact presented scenographically on the court stage – in this way the mythological and allegorical space of the court appeared in a rare example in theatre history as the space of the theatre staging itself.

It may be pointed out that there have always existed theatres of two kinds or orientations: those with exteriors and/or interiors that carry a "message", or even a whole ideological programme independent of whatever performance is staged within their space, and those which do not carry any specific meaning, except for being a permanent or temporary acting area. Its function is the only meaning. Our basic concern in this context is whether and to what extent the building and its interior become part of the performance. This space not only creates the atmosphere and provides the physical conditions for the performance to take place but can also refine, explain, negate and influence the receptive processes. So, one is inclined to ask further questions.[6] What exactly is the relationship between the auditorium and the stage? What is the artistic or ideological programme of the interior or exterior of a theatre? In what way will that programme be corroborated, neutralized or even negated by the spectacle? In what way will it influence our perception? An architectural historian will focus on permanent features of the building, like the structural elements and the relationships between them, the characteristics of the interior design, ornamentation, displayed works of art, affinity to other buildings of the same kind and so on; the theatre historian will take account of all that, but will focus on the temporary significance of all those elements during the actual performance, that is during the time when the theatre interior may become an inseparable part of the spectacle, thus creating yet another network of relationships and an additional layer of signification.

Dynamic stage

In theatre the stage itself is not only a designated playing area but a material attachment or anchoring of the denoted world of fiction. It is always a space of reference, whatever its physical shape might be. It might be a simple rectangular plane or an opened-up body, it might have the most varied of shapes, but it cannot fail to be material. In its simplest form it is two-dimensional (a plane), in a more advanced form it has the features of a three-dimensional body. On this is anchored any kind of fictional space created during the performance, which is usually presented as having continuation in space outside theatre. Within this space, or the retort, there takes place the process of semiosis, embracing everything which is found in its confines. Not only is it a functional plane or body, defining the playing limits, on which you can tread, on which scenery can be built, but the stage itself may be a signifier. The stage can therefore have any shape and it can be designated anywhere, not just in a "real" theatre building. It must, however, be ostensively clarified and defined, even if only by the word and acting; although most frequently it is built up with scenery, namely objects which usually create a metonymy of a larger space, or the illusion of an architectural, indoor or open-air space (there can also be non-representational compositions).

It is therefore important for the fictional space to have at least a minimal but permanent material contact with the stage, for its signifier to be material and visible: we call this the physical attachment of fiction, or its anchoring. Without this material carrier the creation of a fictional time and space beyond the stage proper is not possible at all. For this reason a clear distinction must be made between the material type of two-dimensional stage – for example, a piece of asphalt marked out by chalk on a town square, and the three-dimensional stage – for example, the same piece of asphalt but, instead of the chalk circle, with an upright wall set on one of its sides, forming the "background" to the action. In the first case, the space has an amorphous character, and not only because the audience can stand on all sides but above all because there is no reference point which could give it orientation – we do not know where the "front" is, where the "back", where left and right are (Figure 1). The actor by him/herself is not enough, since he/she is usually not static and cannot constitute a stable point of reference. It is extremely difficult to give spaces of this type the character

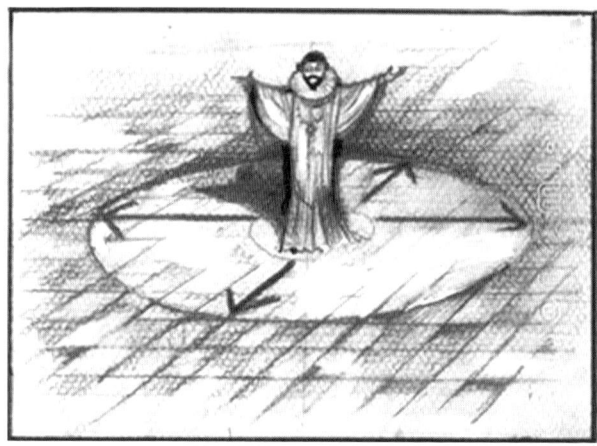

Figure 1   An actor performing on a two-dimensional stage (drawing by Andrzej Markowicz)

of a body, even a fictional one, to establish its dimensions and the internal spatial relations between everything which is found within its confines. In the second case the situation is much clearer – we know where the front is, where the back and even the "sides" are; furthermore the wall is a spatial border between that which belongs to the playing area and that which does not belong to it and which is hidden by the wall (Figure 2). Since we can see that the wall is material and really can conceal something, it is so much easier for us to accept conventional, diegetic spaces (in "real" theatre most spaces talked about or implied are in fact concealed from our sight).

Let us take a look at the problem of spatial relations in theatre from yet another angle. There is no doubt that there is a difference between the implication of fictional time-spaces in the areas of the theatre that are really invisible to the spectator, for example in the areas adjacent to the box stage, and the same implication in the case of the thrust stage, which on three sides at least has nothing to hide. It is true that in the first case, the bodies, the words, light, objects, movement and actions create a fiction, for whose limitations our imagination must compensate, but this fiction is objectively curtained off from our eyes, so in some degree its unreality remains inferred. The concealment of part of the stage from the eyes of the audience provides the justification and rationalization of the effect of stage synaesthesia, that is the creation of a fictional world by the use of senses other than those to which theatre signs refer. This justification and rationality is lacking in a much greater degree in the case of the thrust stage: here there is usually material support on only one side, while on the other three the stage can conceal nothing. Hence the synaesthetic or blending effect will demand

Figure 2   An actor performing on a three-dimensional stage (drawing by Andrzej Markowicz)

greater involvement on the part of the spectator, who must create in his/her mind everything that cannot in reality be seen. The creative effort of the audience will be greater, but by the same token, as we may suppose, so too will be its involvement – also in the sense of co-participation. It seems, however, that in the theatre it is very difficult to create a coherent fictional world without the help of material support to make possible the conventional concealment of fictional spaces. Hence the need for the stage to be supported (at least on one side) by a construction that can cover "something": it may be an ordinary curtain, tapestry or partition, a shed or a hut, or the façade of some building. This also gives support and anchorage to the fictional spaces created on the stage. If the talk is of a street in a town, we understand that behind that curtain (now: the façade of a house) there is some invisible interior, adjacent to a space shown. If the stage is the approach to a fictional castle, we understand and thus more easily accept its complete and visible conventionality, if the curtain becomes a sign of the castle walls, behind which, as we infer, spaces extend that are invisible to us. But invisible in a rational way, because they are curtained off, as we can see. Thus stage synaesthesia is a basic means of creating emergent structures in the mind's reception, and its gradations will vary according to the shape of the stage. It may actually be explained in terms of cognitive or conceptual blending, with the material stage being one of the input spaces, the denoted space or its succession being the other input spaces, and the relationship between the two creating the resultant structure or blend. It is the ability of the human mind to blend conceptual spaces that results in the cognitive rendering of the space that goes far beyond the empirical reality.

The denoted fictional space and the resultant cognitive space in the mind of the spectator "disperse" in various directions (NB usually not in all!), but not with the same range or intensity. Furthermore it is not in its entirety available to our sensory perception – what we see is usually only a part of a whole. In the case of a two-dimensional, and therefore amorphous, stage we cannot, however, define in which directions this space is dispersing. For this we need a materially three-dimensional stage (a metonymy of a cube), and therefore one that is oriented. But only one of them – the space of the three-dimensional stage – is, in contrast to natural spaces, oriented and modelled and may also gain a temporal dimension; that is why there is no need or possibility of defining its azimuth – there is not the slightest danger that someone will get lost in this space. This becomes especially visible in stage compositions constructed on the principles of the linear perspective. But the variants of the geometry of the stage are numerous. Even the fictional implied natural space is oriented and intentionally modelled in such a way as to clarify other elements of the performance in a way desired by its creators. Each change of scene during the performance can cause a change in this orientation (and thus its relation to the external world, its degree of homoeomorphism, will be different). From this arises the considerable elasticity of the fictional space and its changing borders and external points of reference. The denoted space unequally stretches in one, two or three directions, up or down, its frames change, it embraces the auditorium or cuts itself off from it, it maintains continuity or it does not, and sometimes, in extreme case, it becomes closed within itself, that is it is limited to that which can be seen and does not imply that it is a fragment of anything whatsoever; it can also be relativized, twisted, turned upside down, contrary to the laws of physics which govern the Euclidean geometry of the three-dimensional space as the example discussed below shows. One more remark – with reference to what was said earlier – fictional (i.e. scenic) spaces can and should be distinguished as oriented "away from the auditorium" and "towards and through the auditorium". The former I suggest calling escaping or vanishing (just as in perspective the point at which two straight lines meet is escaping or vanishing), the latter incorporating or levelling, since to a certain degree it "incorporates" the space of the auditorium, and levels the invisible barrier dividing it from the stage. A simple aside incorporates the space of the auditorium. This distinction is useful to the extent that in a performance we can have an alternating game of orienting the space. I call such a space pulsating. Sometimes in the scenic space there can be recognized various orientations at the same time. In that case we shall call it simultaneous.

However, in the case of the alternation of the spatial orientation of the performance, i.e. the pulsating space, the picture plane, which creates a material attachment of the escaping or vanishing space, will be necessary; this point of attachment is so important for its structuring in the reception, defining as it does the border between the two types of orientation. The best known example of such a stage is the ancient Greek or Roman theatre or the Elizabethan, or, generally speaking, a stage whose spectators do not surround it from all sides but from two or three, thanks to which the fictional spaces have some material anchoring (for example behind a curtain or the backstage façade). There is also a kind of space which tries to avoid orientation with the exception of what is seen on the stage. We can call this an enclosed or adducent space. In this case the composition of the stage and the performance text do not imply that some world exists on t he outside (many examples may be found in the experimental theatre of the last century). Another atypical example is a staged exteriorization of the thought processes of the figures, for example a dream, memory or reverie. We understand then that the space of the stage – and what is being played out in its material frame – if we do not see the figure – is in essence the interior of the brain and the processes taking place there. The perceptible, material, space is identical with the fictional space, which has no continuation beyond the stage. For example in the 2001 staging of *Hamlet* (under the title *The Lazy Revenger*) by the graduate students of directing at the St Petersburg Theatre Arts Academy (Professor Vadim Golikov's class), the Danish Prince reads ancient tragedies in silence (preparing for a performance) and during his reading there appear on stage appropriately lit flat figures on red material, vividly reminiscent of Ancient Greek vase painting (black-figured). We understand that the scene – which does not appear in Shakespeare's original – is presenting the thought processes going on in the mind of the reading Hamlet. In other words, the space of the stage becomes a sign of the space enclosed in the skull of the figure, transformed into a subjectivized space (as happens in narrated and commented upon scenes of memories from the past, dream insets or interpolations of the future). The material walls of the stage polygon denote the spatial borders of this world (a sign of a skull).

Practically all the typologies of space in the theatre known to me are static in their treatment and above all concern the function and the user; the distinction proposed above introduces new features – dynamics, orientation and a temporal dimension. In order for an escaping or vanishing space to exist, there has to exist – as I have already mentioned – a conventional picture plane (known as the background in the science of perspective), through which we observe it or behind which it remains hidden but exists as a conjectural, implied space, inaccessible sensually. The latter must also have some material attachment or anchoring to the stage. This creates a very significant point of reference, indexality, thanks to which the space of the stage rids itself of its amorphousness in favour of structure. Now we know where the front and the back, left and right, and so on, are. A typical example of such a space is the box stage. In the case of an embracing (levelling) space, there is often no need to establish a point of reference; it always appears when the articulation of the performance text – with at least a partial retention of the space and time separation – includes within its confines the auditorium space. A typical example of such a space is street theatre, or some forms of medieval theatre, that is mainly a stage which is surrounded on all sides by spectators. Apart from the physicality of the performers there is no point of attachment which would enable its spatial structuralization by the recipient. This space, typical of many early theatrical forms or street theatre, is in principle amorphous, and the fundamental point of reference becomes our – the recipients' – point of view.

Theatres and their interiors

As indicated above, it is possible to distinguish theatre buildings and interiors which are independent of the performance, and those that are complementary and integrated with them. Consequently, this enables us to distinguish theatres in which triple relatedness may be observed (the staged spectacle vs. ideology of the theatre and its interior vs. the empirical reality outside) and theatres with dual relatedness (the staged spectacle vs. empirical reality) only. The first of these will usually characterize those theatres which were arranged in buildings that were either occasionally used for theatrical purposes, like Romanesque and Gothic churches, or were part of a more complex architectural structure, like all sorts of banqueting halls and theatres in royal or ducal palaces. Buildings originally designed as theatres quite often reveal a conspicuous ideological programme that can influence the perception of a performance, as is the case of class-oriented edifices, aristocratic, bourgeois or working class. Occasionally these will have an ideological programme revealed in the architecture and design of the interior which carries an educational or propagandist meanings ("temples of culture", "high-society", "people's theatres", "underground theatres" and the like). In practically all of these interiors, the space of the auditorium is separated not only from the artistic realm of the performance, but also from the physical space outside. The iconography of royal or ducal palaces was often a reflection of the then-current ideological programme of a given dynasty, and was usually designed to mark a rigid boundary between the semi-divine and super-human world of the palace and the ordinary human world outside. The space and time/lessness of these interiors, inhabited by superhumans, are often separated from the space and time that govern the mundane lives of ordinary humans. Thus, a performance staged in a space meaningful in itself is bound to create relationships and meanings peculiar to that space and the circumstances of the actual spectacle. In other words, the fictional realm created during the performance will inevitably be related to both the space of the theatre and the world outside it.

   The second type of relatedness, that of a dual nature, will predominantly characterize all those theatres which were originally designed to serve a specifically theatrical function and do not dissociate themselves from the outer reality. In many of those theatres the time and space of the auditorium is part of a larger whole, of empirical reality. Thus, we may distinguish two extremes of theatre spaces. One is oriented towards itself, and allows for a limited variety of performances which will tend to contribute to and complement the meanings of the interior, and, in this sense, this introverted theatre might be labelled autotelic: the primary function of performances staged in that kind of theatre will be to corroborate or elucidate (or both) the meanings generated by the theatre building and its interior.[7] Because the other extreme is semantically neutral, without significance, it is extraverted in character, amphoterous, and oriented towards the outer world and towards an infinite variety of plots and meanings. The first type may be exemplified by some historical cases of court and bourgeois theatre, but above all is manifested by the Christian church-as-theatre; the other kind is evident in all types of "poor" theatres, of which the Spanish *corrale de comedias* and the London theatres of the period between 1567 and 1642 are perhaps the best-known manifestations.

It may be added that most contemporary theatre interiors do not reveal the tendency to acquire meaning during the performance: lights are faded out just before the performance, other lights are lit when the curtain is drawn aside, so that it becomes clear what the director wants us to see, and what he or she wants us to neglect. In other words, lighting strengthens the division between the semiotic sphere and empirical reality, which is blackened out (the reality of the auditorium is faded out into practically non-existence). Moreover, we, the spectators, also find ourselves in the blackened area: this limits our ability to watch others who are not actors on the stage, and in addition brings relief from being watched by fellow-spectators (it does not really matter whether or not the latter is comfortable to us: the visible presence of other people creates a crowd, and, consequently, evokes the feeling of a confined space). Psychologically that means that once the auditorium is darkened, the crowd disappears and our perception of the space opens to greater dimensions, not limited by the presence and sight of others. In this way, the materiality of theatre space is partly blurred or even suspended for the time of performance; partly, because we are still able to see something in the dark, and all sorts of noises, like coughing, laughter, whispering, etc. remind us of the existence of other members of the audience. Thus, the darkened auditorium, along with perspective scenery, creates better conditions for an individual, or egocentric, perception of the spectacle, and marks a clear-cut division between the artistic and empirical realities.

However, the history of indoor theatres with artificial lighting is relatively short, and throughout most of their history theatrical performances took place during daytime, when the amount of light did not divide the stage from the auditorium. What this means is not only that artificial lights were not used on the stage, but that there were no technical means to darken the auditorium.[8] Instead of dematerialized space, we have a clearly visible and measurable space of the interior, which never ceases to be part of the world outside. Instead of an egocentric, individual perception of the performance, we remain part of the crowd and we retain the continuous sensation of communal perception of whatever is presented to us, which among other things is caused by our ability to observe other spectators freely, along with their reactions which – regardless of whether we find them appropriate or inappropriate – influence our own perception and reactions.

Summing up, it can be said that it is the orientation of space and its grounding in time which decide about the creation of fiction and divide the world of performers from the world of recipients. The stage world comes to us, we do not go to it. Thus, the creation of fiction on the stage depends on, among other things, the imparting of dynamism to the space by giving it an orientation which is independent of our point of view, grounding it in a time which is independent of the recipient's time, and the unequal and changeable limiting of its continuity, i.e. modelling. The key role here is played by the actor's body, which may be seen as a biological clock on the stage, which has the ability to disperse or impose its own time upon everything that enters into indexical relations with it. This is why the imposition of the fictional time structures (the effect of theatricality) is not confined to the space of the actor's body, but may include the surrounding phenomenological space. Consequently, there are at least two basic modes of carrying out a theatrical performance: one, when the space of the production is semantically limited to the spaces generated by the actor's body, and various configurations of these bodies, with the spatial material context neutralized or open to the acquisition of any meaning imposed verbally or through acting; and two, when the actor's body is set within a larger meaningful space, adapted or created *ad hoc* for a given production, and thus a more complex network of spatial relations between bodies and objects is established. In the first instance, there is no iconic similarity of the material space surrounding the actor's body to the space denoted by his/her utterances, gaze or gestures; this is why in that case the relationship between the referent and the signifier is always indexical, not iconic. In the second instance, the material space surrounding the actor's body has been visibly modelled or purposefully selected to reveal at least some iconic similarity to the referents; this is why the relationship between the referent and the signifier is not only indexical but also iconic. In the first instance, the signifier is flexible, revealing semantic fluidity, and after serving the function of meanings, say, a façade of a house in London, in the following scene it may reveal its amphoterism and mean a castle wall, or a slope of a mountain. In the second instance, it is not easy for a scenographic slope of a mountain, mimetically represented, to gain a new meaning of a wall of a castle. The set has to be changed, unless the particular production does not follow mimetic rules.

The medieval Christian church, the Renaissance court theatre, the present-day box stage with mimetic set – are all examples of the integrated mode, where the space of the actor's body is incorporated into a larger sceno-graphic space, which is meaningful in the sense that it is in itself a vehicle of information. Following our distinction, it may be said that the space is *integrated* when the phenomenology of the stage contains elements that reveal at least some iconic resemblance to the denoted fictional realm, as clarified by description, acting, etc. When there is no resemblance, and the relationship between the signifiers and the signified is only indexical, signalled solely by the actors' utterances and ostensive gestures, then the mode is *unintegrated*. The Elizabethan theatres may be seen as spaces which do not iconically integrate with the space of the actor's body: it may even be said that it is a space composed almost entirely of the actors' bodies and their different configurations, and the space outside the bodies is denoted verbally or through acting (gaze, gesture, other ostensive signals). In other words, the stage and the theatre interior are an extension of the actor's body and its temporal value in a given scene. Practically all the meanings are created by the actor, whose body should be seen as the Time and Space Generating Component. It is the actor who clarifies what is not seen or heard on the stage. Much of today's theatre reveals this tendency. Between the two extreme poles, there is the whole history of theatre with its amazing spectrum and variety of spaces created or adapted.

Theatre in the house of God

For theatre historians the tenth century is a miraculous period for it gave birth to Medieval drama and theatre.[9] By introducing a theatrical scene into liturgy, the first theatre directors of Medieval Europe did something unprecedented: they evoked a scene from the historical past, from a differ-ent space and from a different time and presented it as if taking place here and now. In other words, they created a fictional realm in which both time and space are conventional. It is indeed remarkable that in a simple dia-logue its theatricality may be detected: for the first time since antiquity the essential feature of the art of the theatre could unveil itself – the empirical present time could blend with the fictional one. On Easter morning, from a church interior we are taken centuries back to a location distant from us by thousands of miles. The separation of the artistic reality on the stage is marked by yet another convention, being the basic rule of theatricality: coexistence of two models of perceiving reality. Figures in the performance communicate with each other (via the actors) and pretend that they do not notice our (recipients') presence. Thus for the first time in history liturgy is enriched by acting. Three Marys visit the Sepulchre and meet an Angel who tells them that Christ has resurrected (Figure 3). With this moment a new phase in theatre history begins.

The interior of the church may be seen as an illustration of both the Bible, the New Testament in particular, and of the human world; and even more

Figure 3   A conjectural image of a liturgical performance (drawing by Andrzej Markowicz)

an illustration of the whole universe, not only because various iconic works are displayed inside, but also because of its meaningful architectural shape. The church as an architectural edifice is an encyclopedia, a compendium of knowledge and cognition of the world, of history and, of course, of God. The level of signification of the interior and exterior is so high that we can actually talk about the church as a text which may be decoded and interpreted. This is the area of research of art historians. Let us just consider the geographical orientation which in this particular case is especially meaningful: the nave and the choir lie on the East–West axis; the entrance is in the West end whereas the altar is located on the opposite side, very close to the East end. Symbolically the East is always connected with light, new life, paradise and also Easter (the etymology of the English word is quite obvious), and consequently with the Cross, standing on the altar. Theologically, East is linked to the Resurrection. West is the worldly side, material, human, whereas East is the divine, heavenly sphere. In addition, through purely architectural articulation of the interior, through columns, half-columns, piers, gallery, triforium, clerestory and the like, not only an internal rhythm is created which plays a constructive and aesthetic function, but also a general direction is determined which leads the worshippers from the West to the East, from the earthly sphere to the heavenly one, from measurable space and time to the immeasurable. This road inevitably leads visitors to the altar, and to the Cross, because it is only through the Cross that we may reach salvation, an eternal union with God. The church architecture and its interior design show us how to reach that goal. The very fusion of the Cross with the altar is meaningful: first altars in Christian churches had the form of a Roman tomb and were inevitably associated with Christ's sepulchre. Symbolically, and theologically, He still remains there in the tabernacle in the form of wine and host. Thus it is through the altar that we reach the Cross, which is not only the sign of Christ's suffering, but predominantly of Resurrection and Salvation. In its architectural shape, the church is also a cross; moreover, it may be treated as a sign of Christ. It is a meaningful structure, theologically significant, and its interior is filled with allegorical and symbolic meanings. Thus, having entered a church which is a cross, we are led to the crossing of the nave with the transept, beyond which we reach the altar, surmounted with the Cross, and there, in the tabernacle we find Christ's Body and Blood. During our lives we can reach the union with God through Communion, but the final and everlasting union comes after our death, when we have finally "shuffled off this mortal coil", delivered from the material world. Thus, it is the church that through its architecture illustrates and shows us the way to that union. In front of our eyes the Cross of Christ's Passion is transformed into a huge space of the cross of the church. In other words – the church has its roots in the cross, but at the same time functions as its illustration, an enlarged projection of the cross's interior. At the same time it is the House of God on earth. It is also a model of Heavenly Jerusalem, where, according to Ezekiel (47.12) a tree is growing in the centre. Entering into the church we are in fact entering into the House of God. We experience Christ's Presence in at least two ways: symbolically, in bread and wine and physically in his iconic presence showing His agony.

Liturgy is an enactment of the miracle of transubstantiation. Starting from around the mid-tenth century, during the Easter liturgy we witness a presentation of the first Christian drama with Three Marys and an Angel. The action is meagre, so are the sung dialogues, constituting the *Quem queritis* trope. But nevertheless the play takes us back to the year 33 and to a different, historical space. We drift in time and space as in a real theatre. But this is not simply a play, but a dramatic piece inserted into a larger text of liturgy or the Church itself. Liturgical play takes place in Jerusalem or near Jerusalem. One should add – in earthly, historical Jerusalem. At the same time we must not forget that the stage of the play is located near or around the altar in the House of God, in a small area inside a model of the Heavenly Jerusalem. A scene from Holy Week is presented, culminating in the Resurrection. And the Cross is the sign of Resurrection. Thus, the earthly sphere is contrasted with the Heavenly one (the space of the church – through consecration – is separated from the empirical reality). The historical Jerusalem is shown as a process of transformation from the earthly city, inhabited by people full of vices and virtues, living the lives in a human time, into a timeless Heavenly Jerusalem, inhabited by the souls of the Saved. This is the road to Salvation. Thus, what we have in this short liturgical play is not only a scene standing as evidence for the miracle of Resurrection, but also the mystery of the Cross. What is shown is the transformation of the earthly into Heavenly.

We may therefore ask the question: what actually happens to the Cross and the figure of Christ during the performance of a liturgical play? Well, first of all, the crucifix is covered with a layer of fictionality. This will result in a change of semiotic significance: the Cross will cease to be a Christian symbol only, because it will also become a theatre sign. A sign of what, one may ask? Well, during the performance the cross becomes a sign of the historical, material piece of wood on which Christ was crucified. Thus, the duality of the cross will be fully revealed and the effect of theatricality will mark its presence. The same happens to the figure of Christ. During the performance the figure is taken down from the cross and laid in the tomb; it thus ceases to be a symbol only, becoming a theatre sign – one of the figures in the play, i.e. the historical Christ who died on the Cross. Hence, the duality of the figure of Christ is foregrounded. Let me stress that whenever the fictional layer of time and space is superimposed on an object which in the empirical world plays the function of a symbol, this function does not disappear altogether, but is simply weakened. Owing to the fact that both layers intermingle with each other, we can observe the phenomenon of "twinkling" of meaning. With the end of the performance, all its elements are deprived of their historicity and fictionality. Everything returns to the here and now of the church or the timelessness and eternity of the symbol. Having played its role as a figure, Christ returns to the cross, losing His features as a theatre sign, returning to His significance as a sign of everything that the Cross means. Once again, the earthly becomes Heavenly. Thus, the performance itself shows us the process of transformation of the material into the immaterial.

The complexity of theatrical signs in liturgical drama is caused by the unique time and space of the church interior. We can even say that the physical time as we know it does not exist there (of course, we, being worshippers or visitors entering the church bring our individual measurable time); the sacred space is timeless and may be linked with incommensurable infinity rather than with physical space. Thus, quite contrary to other types of theatres, the church interior, being a model of Heavenly Jerusalem, is not part of any external space, or even human space. This explains why, during the actual performance of liturgical plays, the artistic temporal–spatial continuum will create as it were tension between itself and two essentially different spaces and times, the one belonging to the empirical reality and the other sacred. And that particular feature makes liturgical spectacles distinct from any other theatrical forms. Similarly, all theatrical signs created during the performance will reveal a threefold structure: the usual fictional layer will be coupled, so to say, with the layer of theological meaning. Thus, to give an example of the complexity of signification, the Cross from the altar when used for theatrical purposes will become a stage-prop before the actual performance, but during the performance will become a sign of the original Rood, the historical material cross on which Christ was crucified, but will also remain its theological symbolic significance. In other words, will reveal its triple nature: theatrical, historical and eternal. The same will occur in all other instances involving those elements of the performance, which, in the non-performance reality have a theological, symbolic or allegorical meaning. Thus, the keys used during the performance by Peter, in a later, more developed liturgical play, will be a sign of both material and historical keys belonging to historical Peter, one of the followers of Christ, and of keys being the attribute of Peter the Gatekeeper of Heaven. Resulting from this will be the effect of the alternating dominance of historical time and space, fictional time and space, and theological eternity. All three will co-exist during the performance rather than annul each other. All the remaining elements of the production which do not fulfill the above mentioned functions (stage properties, costumes, actors, etc.) will become constructive elements of the production, creating a different historical time and space. Consequently, the unique co-existence of three different times and spaces will result in a tension between them; moreover, will lead to the conclusion that liturgical drama cannot be separated from liturgy without seriously undermining its unique features. As we know, further development of dramatic forms leads towards greater independence from liturgy, which inevitably ends in a total separation from the church.

The wooden O

Let us now turn to the other extreme type of the theatre interior, which in itself does not have any meaning other than functional, of which the *Globe* is perhaps the best known example.[10] In other words, the space of the *Globe* does not mean anything apart from conveying the following message to us: I am a space where plays are performed and watched. From this perspective, Shakespeare's image of the "wooden O" (as he calls his theatre in the Prologue to *Henry V*) would not only mark the circularity of the frame of the theatre, but could also mean "zero", the "crooked figure" that denotes a space that before and after performance signifies nothing, apart from being a performance space. It is only during a performance that certain structural elements of the building begin, through a layer of fictionality brought by the actors, to mean something, and for this reason may contribute to the infinite varieties of meanings generated by the performance. It is the fictional realm created by the actors during the actual performance that makes the otherwise semantically neutral interior of the theatre mean something. Owing to that feature, the variety of possible meanings and the ease with which they could be created is truly amazing. Thus the "cockpit" turns into the "vasty fields of France", and the "unworthy scaffolds" into the globe. The gallery in the tiring-house façade may become a balcony or the crenelations of a castle. The "heavens" above the stage may become astronomical heavens, by which the stage becomes the earth or the globe, creating thus a model of the universe. Changes of space are concomitant to the jumps in time. Thus, it may be said that these theatre interiors become extensions of the actor's body and time – which have the ability to create fictional present time, it must be added. However, all of this happens during the performance only: before and after the stage has only one meaning, that of a wooden, raised platform which is used for dramatic presentations. The frame means three storeys of galleries for spectators, made of timber and very simple in their construction (Figure 4). However, owing to the fact that neither the exterior nor the interior of that type of theatre means anything except an area for performing plays, during the actual performance of a play, any physical element of the theatre may become literally anything, depending on the artistic demands of the stage text. To see this theatre as a philosophical construct and a sophisticated model of the universe would be to impose a meaning that may occur during a brief scene in a play but does

Figure 4 A conjectural image of the stage at the Globe (drawing by Andrzej Markowicz)

not exist at all before and after the performance. In point of fact, it would seriously undermine the particular nature of these theatres.

One of the distinctive features of the Elizabethan theatres is their stage's uncommon ability to transform into practically any space in a blink of an eye. This allows frequent jumps in time, which add an almost cinematic quality to Elizabethan stagings. The phenomenon is worth at least some attention, because finding the answer will also show us the ways in which the communication vessels work in theatre. So, how does it happen that in the Elizabethan theatre the stage maintains such great signifying mobility and in the twinkling of an eye can rid itself – like a sponge squeezed out – of its absorbed and accumulated meanings, in order to transform itself, say, from "the deck of a ship" to "the tiled floor of a church"? It seems that this phenomenon can be explained by the fact that everything to be found on the stage has a "common (material) denominator", which for the spectators is the perfectly visible, raised platform stage. Everything is anchored in this. However, because the meanings of the thrust stage are created by the actors only, mainly by words, gestures and gaze, it is also words, gestures and gaze that decide changes to these meanings. This requires no great technical feats: all that is necessary is to clear the stage of actors and objects, that is to annul its semantic value (in radio theatre, silence has a similar function). On the box stage, this is a matter of great difficulty where both actors and objects are concerned, because changes in the stage's meanings are brought about principally by changes of scenery (often combined with dimming of lights or a temporary fall of the curtain). This is sometimes a complicated technical operation, impossible to perform in an instant. Besides this, changes of scenery cannot be made as frequently as changes created verbally.[11] The verbosity of Elizabethan texts becomes more conspicuous when staged within elaborate and mimetic scenery, where things talked about are actually seen. Originally, they were talked about *because* they were not seen, and were to be imagined; within the set they become pleonasmic.

The thrust stage, in turn, having no larger scenery construction, still less any perspective, is ready for constant changes of its meanings, signalled mainly by the actors and verbally. Of course it absorbs meanings, thanks to which we know that we are in the queen's (Gertrude's) bedchamber at the castle in Elsinore (if it is a bedchamber), in which there is a bed (even if we cannot see it), and also a tapestry, behind which a person can hide (Polonius). The stage forms not a frame, but a base, a game board for the episode being played out, and from this base, through meaning implemented by the actors, it grows out beyond itself, hence the queen's chamber may become larger than the materially measurable dimensions of the platform. In other words, the platform of the stage is not the material indicator of the extent of space, but the metonymic sign of space that may stretch out over the whole theatre and beyond. Not being an iconic sign, the stage's meaning is totally dependant on the actor's words, gaze and the props. But at the moment when all the actors exit (and all the objects are removed),[12] the stage is immediately reduced to semantic zero, it ceases to by a metonym of anything (except of the theatre): the actors take out its meanings and it is ready to receive new ones, brought in by the new actors who enter (and the objects brought on). That is why it is so important to clear the stage of all the objects and actors that took on the spatio-temporal meanings of "the queen's bedchamber", if we are to move from one scene to another taking place somewhere else and at a different time. If it is not cleared, a single object will, metonimically, signify the space and time continuity from the previous scene. That is why it is so important for Hamlet to drag, for no apparent reason, the body of Polonius off the stage, or for Edgar to do the same with the dead Oswald in *King Lear*.[13] Interestingly enough, when the body or an object is first taken out, along with everything else, and then brought back on the stage in the following scene, it has already been cleansed of the former attributes of space and time, and gains new ones, appropriate to the new scene. We find a scene like that, to give one example, in John Webster's *The Duchess of Malfi*, where in scene 4 act 5, Antonio is being murdered and his body taken off the stage, to be brought back in scene 5, which takes place in a different location (though still in the Cardinal's house). Because changes of scenery are not required to change the meaning of the stage, it can be subject to such changes repeatedly and literally in the twinkling of an eye.[14] The only condition – let us repeat once more – is its reduction to semantic zero. Without this, each new figure that enters is ascribed to the time and space presented so far. Hence the Elizabethan stage emerges not as an iconic "picture", but as a game board, characterized by its ability to make constant spatio-temporal changes, brought into being verbally and by the sheer presence of the actors. It is a semiotic chameleon, which, while colourless to begin with, immediately takes on colour when a new stage figure steps on to it. After the stage has been cleared of actors and objects, our chameleon instantly loses this colour, becoming blank once again, but only to take on another colour immediately with the entry of another actor. And so on. And as Hamlet says, the chameleon lives on air, just as the theatre for which the prince himself was written, lived on words, whose substance, as everyone knows, is air – nothing, that is, zero to appearances.

Another remarkable feature of theatre stages is that different fictional spaces can be denoted on them simultaneously. Let me give just one example from William Shakespeare's play. In scene three act five of *Richard III*, the king, having reached Bosworth Field, orders: "Up with my tent!" in line 7, and makes an exit (with his followers) at line 18. At the other door enters Richmond, and from the dialogue it is apparent that another tent is being pitched on the opposite side of the stage, for Richmond invites some of the lords inside: "In to my tent, the dew is raw and cold" (46). This allows a simultaneous presentation of events and dialogues occurring in the two camps, without the need to clear the stage of figures and properties.[15] Thus we have two permanent locations signalled, the two tents, and everything that takes place immediately in front of them, with actors making their entrances and exits from and into the tents (and also through separate doors), is understood as taking place in permanent and clearly defined spaces, the dimensions of which might be similar to those in the denoted world. As a result, a new network of relationships can be achieved, for whatever is said on the one side is juxtaposed with dialogues on the other. Also, the peaceful sleep of Richmond is contrasted with the troubled dreams and nightmares of Richard (which are scenically presented!). Secondly, the area between the tents becomes the sign of the battlefield, as the succession of "excursions" and "alarums" shows. Thus the middle part of the space of the stage becomes a metonymy of a larger whole, and its semantics will be different from that of the area around the tents. In each case, when all the figures leave the stage through the door, and another set enters (through another door), the implication is that we are still on the battlefield, but simply on a different part of it. The convention worked even if no material elements of the stage set were used. Simply, during the battle, the exits and entrances of armed men were treated as if taking place within the same field, only in a different part of it. In other words, during the battle scenes the spectator knew that unless a different location was specifically indicated, the location was roughly the same, within the implied field. Accordingly, we have numerous plays (e.g. *Troilus and Cressida, Henry V, Julius Caesar*), where no scenery is used, and the convention is applied in the way I have described. In today's theatre practice, one does not even need objects (tents) to mark different spaces – these may be marked by a different light. In fact, if two actors' bodies share the same source of light, this could imply that they also share the same space. This is why the differentiation of light (colour and intensity) makes it easier to mark the spatial distance. And this was a convention specific for the staging of battle scenes, as is evidenced by the extant plays. So, then, the clearing of the stage does not erase the initial meaning of a battlefield (with the tents still standing there), and the entrance of a new set of actors does not mean a totally new location: it is a different area, but retains the general meaning of a battlefield. It could mean a new location outside the battlefield only if additional information were supplied.

What was impossible in the frame of the perspective stage (or in film), here becomes an acceptable convention. In addition, the stage can function like a kind of semiotic projector, whose meanings emanate beyond its material surface, if necessary filling the whole theatre interior. This is possible, let us add, thanks to the large signifying mobility of the stage, its amphoterism, and also thanks to the absence of a frame and of perspective, which would impose a spatio-temporal unity on the stage, restricting its expansion. During the performance, the thrust stage has no clear demarcation of its boundaries and no unchanging material support indicating its limits. It is a piece of modelling clay. Even the theatre building does not restrict it in going "outside itself"; it does not determine its shape. On the contrary, it is the stage that decides the theatre's meanings; it is the semiotic indicator for the building. Thanks to this the stage – if required – may transfer meanings on to the whole space of the theatre. But the condition for the acquisition of new meanings is the stage's semantic neutralization on each occasion. Thus, theatres of the described type may be seen as an extension of the actor's body, in the spatial and temporal senses, and their initial indeterminacy is clarified by acting and stage speech.

Beyond Euclid

I have already indicated that for theatre studies perhaps the most interesting are the ways of modelling the space of the stage. The degree of complexity of building up the space and of the spatial relations on the stage may at times be quite considerable. Again, this has been described in hundreds of books and articles, so I have decided to limit my discussion to one, albeit very unusual example, which shows that the scenic space is, indeed, like a modelling clay. For in theatre, sometimes the (Euclidean) three-dimensional geometry may be broken down, even on a stage as stable as a box stage, which is in essence an irregular polyhedron. It is beyond doubt that most often the space of the stage is composed in such a way as to create a picture subordinated to the point of view of the recipient. The intention is for the recipient to see and hear as well as possible, and only that which the creators of the performance want him or her to see and hear. At the same time he or she should not necessarily notice that this is an image of someone's view of reality but treat it as his or her own – that is why, in the arrangement of the spatial composition of the stage, it is the recipient's position and direction of viewing that are taken into account, not those of the performers (this is one of the reasons why the actors do not play with their backs to the audience and why we are not shown the "other side"). It must be added that the "real" or factual shape of the fictional space beyond the fifth wall is in a sense an indeterminate imaginary visual composition, and, since the spectator does not see it, it is the task of the creators of the performance to supply such stimuli as to enable him or her to create in the mental perception the desired image of this space. Its function is to create a context, and therefore a further definition and clarification of everything that can be seen and heard, and what is only implied. And one more thing: quite contrary to film, where many shots are taken from a subjective point of view (as seen by one or more of the figures), the space of the stage usually presents itself not as a projection of somebody's vision, but as an objective material object. It is the spectators who view the stage from their individual points of view, but what they see does not want to be interpreted as the stage perceived by someone else, but as part of reality.

However, it is possible – and this is my example – to build the stage picture in such a way as for it to become – just as in a film – a legible articulation of the point of view of the figure. In the afore-mentioned St Petersburg production of *Hamlet*, the scene with the Gravedigger was played in this atypical way and may serve as an example of spatial mod-elling or sculpting that generates meaning. The spatial arrangement was quite unusual.[16] On the back wall of the stage there appeared two separately lit rectangular niches, with their narrower sides on the floor (Figure 5). The one on the left had a dark background like the earth, while the one on the right was as light as the sky. These comparisons are not without foundation, since after a while it turned out that the left niche was in essence the inside of the grave seen from above, while the right one in turn was a perspective of a fragment of space seen from within the grave. How was this achieved? In such a way that the Gravedigger was also seen as if from above: the actor lay on the stage on his back (with his head towards the audience) near the back wall of the niche, with his legs bent

Figure 5 The Gravedigger (Andrei Rostovsky), working with his spade, and Ophelia (Tatiana Zaharova) are shown as if seen from above (reproduced by kind permission of the St Petersburg Theatre Arts Academy)

Figure 6   The two niches are shown as seen from two distinct points of view: on the left, there is Laertes (Maxim Mihailov) and Ophelia (Tatiana Zaharova), as seen from above; on the right, there are several characters peeping into the grave, as seen from below (reproduced by kind permission of the St Petersburg Theatre Arts Academy)

and with his feet against the wall. He waved his spade in such a way as to scoop out the earth (totally conventionally, signalled by the movements of the spade) from the wall-grave interior and throw it to the sides of the wall, i.e. outside the niche. This gave the effect of a considerably distorted, "shortened", perspective, just as if we were watching him from above (Figure 6). The figures appearing from the right looked into the middle of the grave in such a way as to create the impression of people who from various sides were leaning over the grave being dug (only their heads and the top halves of their bodies were visible). Thus the Gravedigger was seen from the point of view of the figures leaning over the grave, while they in turn from the point of view of the Gravedigger. The change of the point of view and the perspective, so typical of film, is very difficult to realize in the theatre, but their simultaneity is possible. Moving our eyes from one niche to another, we not so much change our own point of view (because this remains without change) but receive in turn two different ones (the third one is of course the overall perspective of the director). By accepting and sharing the point of view and perspective of the fictional figures, we experience, at least partly, a rare sensation of seeing the world as if through somebody else's eyes. What makes the experience even more unique is the fact that this "somebody else" is a fictional figure.

What we are dealing here with are two distinct perspectives deriving from two distinct points of view presented simultaneously: the one from within the grave (looking upward), and the other from the outside (looking downward). In the combined geometry of the stage, the back of the stage is both "down" and "up" as if in an open box whose transparent lid shows the owner peeping inside. Consequently, when the actors look straight at the audience, we understand that the Grave-digger on the left is, in his geometry, looking "up", while the others on the right are looking "down" (i.e. into the grave). The curved geometry of the stage is corroborated by other details of the staging, as, for example, by the way Ophelia is laid into the grave. It is also worth noticing that the time of both perspectives is the same and, moreover, flows with the same tempo as the time of the audience. Thus we, the spectators, share the flow of time with the stage figures, and see two picture planes as if through their eyes. The difference, of course, is that we also see *both* picture planes simultaneously, which has a metatheatrical effect. We are reminded that this is, in fact, an ingenious device employed by the director, and the effect of perceiving reality as if through the figures' eyes is only an illusion. In addition, the illusion is invalidated by the fact that the substance of the signalling matter does not change. We still see a theatre stage, lights, actors, costumes, etc. What we do not see, in substantial sense, is the world in the shape perceived by fictional figures. However, we see the "geometry" of their world, its frame, as if perceived by the figures from the two different points of view. This alone makes the scenic device original and unusual in theatre practice; unusual is the subjectivity of the geometry and perspective of viewing which gain a conspicuous indexical function and relationship not to an abstract and invisible "creator", but to the fictional figures represented on the stage by live actors.

The consequences of this are manifold, but the spectator is above all surprised by the breakdown of everything he or she has been used to so far – losing the discernment of where the top and bottom are and of what is vertical and what horizontal. When the Gravedigger throws Hamlet the skull, it flies horizontally from the "window" of one niche to the other; we, however, understand that the skull is not flying horizontally, but in a vertical plane, from the inside of the grave to the outside (obviously in accordance with the laws of the physics of space, since the skull, attached to a string whose other end was fixed to the wall between the niches, flew in a horizontal arc parallel to the plane of the stage floor). This means that we have a simultaneous division of space: in relation to the material floor of the stage (being at that moment a sign of the wall of the grave) both niches are rotated through ninety degrees, while in relation to each other through a hundred and eighty degrees. This literally turns upside down the spatial orientation of the stage, and also its semantics, causing in principle an almost total dematerialization (it might seem) of permanent elements – the floor ceases to be the floor and the support of anything, just like the ceiling and the side walls, which cease to perform the function of the side walls of the polyhedron of the stage (they can only be the side walls of the grave!). The polyhedron in effect is no longer there. That which moments earlier had been the top, side or bottom has ceased to be the top, side or bottom, as in a Cubist painting or rather in a lithograph by Mauritz Cornelis Escher. Thus, when Ophelia, wrapped in a shroud, is laid in the grave – in the understanding of the laws of physics of the world beyond the stage – she lies, standing (Figure 7). The space of the stage once again assumes a certain concreteness only when Laertes, then Hamlet, jumps into the grave. Their fight takes place not only in the narrow niche (obviously, they fight in a vertical position, but parallel to the "lying" body of Ophelia and perpendicular to the Gravedigger, through which we can surmise that in physical reality they are fighting lying down), where the girl's body becomes a victim of their

Figure 7   Ophelia (Tatiana Zaharova) shown in the grave (the actress is standing, but the impression is that she is lying in the grave, as seen from above) (photograph by W. Jakubowski)

struggle – her nakedness is visible beneath the profaned shroud, but also on the main floor of the stage (still parallel to the "lying" Ophelia, but we understand now that their struggle is taking place in an upright position). This further complicates the already muddled spatial relations. Because the laws of physics, optics and perspective have been violated (the Gravedigger is still seen from above, while the young men – because they are standing – are seen from the side), we understand that the floor of the stage in the world of fiction is in essence the bottom of the grave (and not – as before – its wall). This causes a certain inconsistency, since in relation to the Gravedigger, whose position on the floor of the stage establishes its geometry, the young men ought to be fighting lying down. We understand, however, that they have been conventionally "led out" (on the principle of a projection or exteriorization) of the grave interior onto the stage, but *de facto* they remain where they were. Once again the cube – this time the grave – falls apart.

From this example it can clearly be seen that the picture of the stage does not have to articulate always the seeming or intended point of view of the recipient. In principle – in accordance with what I have been saying – it never does this, since it is a concealed projection of someone's point of view, someone's perspective, whose picture, projected on the stage, is adapted to the position of the auditorium, thanks to which we have the subjective impression that it is we who are perceiving a fragment of empirical reality and not that someone is showing it to us by presenting us with a subjectivized fragment to watch, as a specific projection (usually, in mimetic performances, the events on the stage are shown as an objective fragment of reality, and not of someone's thought process). In the scene described above, however, matters become complicated to the extent that the stage picture is not a projection of just one, external point of view (that of the director), which in addition is not always legible, but also the point of view of several figures, their "intra-stage" perspective, which means that – in accordance with what I said earlier – the space does not have one but two (or even three) opposing orientations. This leads to a dismembering of the stage space as a geometric body, and thus to a considerable relativization of it, a twisting and a mixing, and a breakdown of the orientations of its dispersal. For these reasons I propose to call this type of space polymorphic. In fact, what holds this space together is time, not geometry. This example shows clearly that the scenic space can differ greatly from that known to us from everyday experience (and from theatre tradition).[17]

We see also that the orientation and dismembering of the scenic space can sometimes be extraordinarily complicated, and artistically fruitful. It evokes metaphoric interpretations, since in the world of *Hamlet* human relations are not simple, straightforward and predictable. For the inhabitants of Elsinore the "geometry" of human relations is far from being normal; it is indeed "out of joint". One can never be certain as to which direction one ought to choose, one does not know what and who is in front or behind, in both temporal and spatial terms. The introduction of non-Euclidean geometry on the stage foregrounds the mischievousness and perilousness of the space of human relations at the Danish court, where not only time, but also space is out of joint. Again, this meaning is generated through the relationship between the fictional and the phenomenal, which at the same time draws our attention to the rules of the theatre, by which also the aesthetic function is foregrounded. In this way geometry becomes a sign and metaphor of human relations.

5
Sculpting the Time, or the Magical Binder

The evolving present

For reasons given in the previous chapters, we may treat the theatre stage as an almost miraculous space, a translucent chemical retort, in which – during the process of semiosis – the physical laws of our world are suspended and everything that can be seen or heard is engaged in all sorts of reactions and, consequently, is converted into specific signifiers of a fictional world that, in point of fact, cannot be seen or heard. Semiosis needs a binder for a multitude of chemical "reactions" of which every production is composed, and it seems that the special kind of time, unknown in the world outside theatre, that appears on the stage plays that function.[1] This explains why the director's task is "to sculpt the time".[2] Naturally, time does not exist in any material sense, and in its cognition humans employ metaphors, usually linked to the movement through space. Generally, it may be said that time is a relationship between events (consciousness of the act of perception of the world around us can also be treated as an "event"). Since in theatre the occurring events are preordained, so are the temporal relations between them. In theatre there is no future as we know it, there is no past as we know it, there is only continuous, evolving present. This has significant consequences when we consider the differences between the two worlds.

It becomes apparent that theatre arises from the appearance of discrepant time structures and let me repeat: the performer as a live actor, living in the same time as the spectator, "pretends" that he/she is someone from the past (or more rarely, from the future), but behaves as if the fictional figure were living now, in congruence with the biological time of the actor and the spectator. Consequently, time in performance becomes split: it is the real present of the performer (the same as the time of the spectator), and the created fictional present time of the figure. However, both of these present times – while appearing in the shared present time of the audience – belong to different streams, which reveal different characteristics. Without that factor no theatre is possible, although an imposition of physical time structures on the fictional time may occasionally occur; consequently, both streams flow with the same tempo, as is often the case in current productions. But this does not mean they are of the same temporal "quality". What is submerged in the fictional time cannot be presented directly on the stage; the "dissolved" components can only find their material "shadows" or signifiers there. The described duality is perhaps the most important feature of stage time. The fictional time stream is flexible, elastic, and may change its pace even within a single scene. Moreover, the actor signals that presented fragments from somebody else's life are not taking place here, in the space of the stage, but in another space, and does not recognize the theatrical nature of everything that is being shown.[3] This creates a further contradiction, since in empirical reality we cannot have two present times, each flowing within a different time stream. And the human biological time cannot be split into two present times, for the biological present time is a continuum, immune to objective ellipsis. But theatre is not the empirical reality known to us from everyday experience, in which everything has to succumb to the laws of biology or physics.

To make things even more complicated, sometimes we may notice the presence of several different time streams operating on the stage, involving different figures, who, as the result, cannot see each other even though they are present in the same space (as in Thornton Wilder's *The Skin of Our Teeth* or in Tom Stoppard's *Arcadia*, discussed in some detail below). This means that we can have more than two present times presented on the stage simultaneously, the "real" one (of the actors and spectators) and two or more fictional times that flow on the stage parallel to each other. In the Stoppard play, it is the figures from the early nineteenth century and from the end of the twentieth century that share the same space and even some of the props, but yet cannot notice each other's presence. On rare occasions, we may even witness a dialogue between two or more figures, each set within a different time: this usually implies that, with the exception of magic or other supernatural circumstances, we are dealing with a dream vision, imagination at work or hallucination. In a play by a young Polish playwright, Małgorzata Sikorska-Miszczuk, entitled *The Suitcase* (Pol. *Walizka*), directed by Piotr Kruszczyński (*Teatr Polski*, Poznań, 2009), in a moving scene a grown-up son, who long after the war visits a holocaust museum do discover his father's suitcase on display there, converses with his father, who is one of the victims in a gas chamber in a German concentration-camp (the confinement and the agony is presented as a mime show). In order to bring logic into this scene, the spectator is led to interpret it as the result of the son's imagination.

It is obvious, even to an uninitiated person, that every performance has its beginning and an end, and its time does not stretch beyond these limits. There is no past before the beginning, and no future beyond the end.[4] The best way to look at it is as an evolving present, and from this point of view, a performance as a work of art, or events, is composed of the present time only. It only needs time to present its own entirety. Of course, the plot is organized along a different time structure and time flow, and it is a succession of events, actions and utterances arranged along a temporal axis. And in the latter sense events may be said to occur earlier or later than other events; but the performance as a whole does not have its past time, nor its future: it is contained in oneness in a structure that is basically atemporal. This is why the structural elements and components of the whole are not really separated by time or space: they are linked with one another through a network of relationships that are contemporaneous in the sense that they exist in the present. The future in a theatrical performance is not a possibility, as in real life – it is preordained and given. These relationships are unfolded in time before the spectators, but in cognitive perception take the form of a blend, which in itself is timeless and spaceless (although it may be modified in the course of time). Time is the binder that holds the structure together, and draws our attention to various relationships within it. It also marks the boundary between the theatre and the world outside.

Elastic time

Since the events concerning the fictional figure are presented within the actor's present time, it is logical that the same rule applies by analogy to the past and the future. The present always results from the future, and since the former is shared by the figures, actors and spectators, so, by implication, is the latter. And the past is the present passed. Thus, the spectator may fall under the illusion that fiction is set within the same time stream as reality, which helps him/her to "live in the blend". This logic fails, however, when the spectator becomes conscious of the fact of acting and knows that the actor on stage is not the figure enacted. This means that the agreement of the temporal present is only conventional; consequently, the shared future (and past) is conventional too. The contradictions, described above, give rise to our, the spectators', awareness of the different space and time presented in our *hic at nunc*. So, the time to come (broadly speaking – the future) is not the same for performers, the figures and the spectators. Time during a performance is finite, preordained, and within time limits, and within the space of the stage everything floats in the temporal binder.

Of course the figures are not aware of that and the actors – as part of their impersonation – behave as if they did not know the exact future, for in the world of the figures it is not part of the script (as it is to the actors). What is happening to them and what they are saying is part of the unknown, the stream of time that we label the future. It is only the actors who know what is going to happen and what their interlocutors are going to say. The actors only signal that they do not know the future, or that they are surprised by their partner's utterances or behaviour. If they reveal their true awareness, the result is similar to the one Shakespeare described in the mechanicals' performance at court in *Midsummer Night's Dream*. So, it is important to note that the conventional theatrical present time inevitably blends the time to come, i.e. the future, which becomes a seemingly shared time of the fictional figures and the spectators. We must not forget, however, that the fictional future has a marked end ("the last syllable of time", to use Shakespeare's words), and everything that is being said and done on the stage is subordinated to the temporal limits of the performance.

Consequently, the fictional figure as enacted by the actor does not and cannot reveal the actor's awareness of his/her past and future: the actor knows perfectly well what is going to happen and what his/her interlocutors are going to say and what they have already said (even in the scenes, in which a given actor does not take part); he/she also knows what he/she is supposed to do and say. If the latter is forgotten, or the former is revealed, the basic rules of the theatre are unveiled and undermined. So, another convention is constituted: namely, that the actor is not supposed to betray his/her awareness of the events to come and words to be uttered in the fictional realm. Owing to that factor, the actor signals constantly that the figure treats the future of the fictional realm as we treat our future – as only to some extent predictable, but basically unknown, a possibility rather than certainty. The illusion is that we, the spectators, and the fictional figures, share the same future time to come. This enables the actor to signal the fictional present (from the world behind the fifth wall) as if it occurred in our real temporal present, and at the same time signal the fictional nature of everything that is being enacted and shown within the split present times. And his/her biological time is congruous with the time of the audience, that is, they share the temporal present and the future. All of this means that also the future time is split, and the duality of the present is concomitant to the duality of the future. The latter feature appears also in productions that pretend to be a slice of life, which is common in today's theatre: no matter how the show is presented, it always reaches its end, which means that during its presentation the shared future time was only a convention. The phenomenon can be labelled *shared* or *conventional future*. The latter, like the fictional present time, is flexible and can change its pace or appear suddenly through the implementation of temporal ellipsis; and it does not stretch beyond a given end.

Admittedly each performance appears before the spectator in a time defined by the clock. However, it is also as a whole both a spatial and a temporal sign. The former rests on the composition of all components within a single "picture", when all the information is presented simultaneously (alluding, say, to a famous painting, or creating an original image of its own); the latter is the sequence of stage pictures, evolving one into another, which inevitably demands time. But every performance also has a different time inscribed in its constituent structures, for which it is a significant "binder". This structured fictional time of a theatrical work can understandably be similar to physical time, but it is not identical to it – it is rather reminiscent of time in the thought processes of a human being, to the way memory works, or imagination or dreams. Its tempo is different and variable and its continuity is usually disrupted. That is why the temporal structures of a performance have to be constructed in such a way that the recipient can on their basis decode and link the whole in a temporal sense also (for example, as a cause-and-effect series of events). Since time is an important element of the conceptual schemata according to which we perceive and understand the world (which is culturally variable), it is to some extent natural that the performance, as a fragment of reality will also be comprehended and regarded through these schemata. This leads to a confrontation between the temporal structures of the theatrical work and the physical and mental time of a human being. This confrontation is something assumed by the creators of the work and is one of the factors modelling its scenic shape, and also influencing its perception. It seems here that it is the specificity of a person's thought processes which causes him or her to be able to perceive events enclosed in temporal structures – sometimes considerably different from those which we are given in everyday experience – as elements of a homogeneous plot or history (in productions where these are relevant).

Let us also observe that the notion of memory comes into focus when discussing the temporal complexities of theatre. One is the general memory of the individual spectators, of which the memory of the performance is only a part. The latter is of course essential to the proper perception and understanding of the work. One of the characteristic features of the "memory of the performance" is that it is confined in time – as I have already pointed out, it has a marked beginning and a marked ending, and cannot stretch beyond these points. It is the memory of an artistic structure, which marks its separateness from the non-structured memory of our lives (of which it becomes part as an aesthetic and intellectual experience). Our memory, although "edited", consciously or not, is not structured as an artistic work, although it may and does provide imaginary and inspirational grounds for artistic creation. The spectators' memory naturally differs from the implied memory of the stage figures, for the figures do not memorize the world around them as an artistic structure, with restricted time and space limits. It may therefore be said that although in theatre a performance has to unfold in time, and is of specific length, the resultant cognitive structure or the memorial blend is timeless, which does not mean that it cannot be modified in the course of time. Memory as such is a process rather than a permanent state of mind.

For all those reasons the scenic time may be seen as a universal binder of all material substances of which a performance is composed. It enables not only the flow of events, the continuity of everything that we hear and see, but also it makes possible all sorts of reactions between particular components. Without it those reactions cannot take place. In other words, time makes the chemistry of theatre and its cognition work. This is why I consider time to play the decisive role in what happens on the stage, similarly perhaps to water in chemistry, biology and life. But, as I said, this is not a simple issue, for the stage binder is of a peculiar and extremely complicated nature. It does not have a fixed formula, available to all directors, and being immaterial it cannot be described or photographed as other substances can. It is treated and understood differently by both physicists and philosophers; in theatre it gains yet another dimension, hence I refer to the theatre's fifth dimension, and has to be described in a different sense.

As I have already mentioned, time in the performance will always be different from "real" time, its tempo will be different (as in old Japanese clocks, in which the actual length of hours varied in the course of the year, depending on the season), it will usually be more "condensed", so that in the course of two or three hours one can tell the history of a life, even a generation or a couple of centuries. It can also be – albeit more rarely – "extended", or even become a whirlpool. The problem depends on the fact that although the performance refers us to some – more or less concrete – time (but not to the temporal frames of the performance itself as a real event), through which it becomes a kind of concept, this cannot be materially "exhibited" on stage. Time does not have a material substance ascribed to itself, there is no universal icon or emblem of time,[5] but it can be a feature of a given semiotic system (for example, poetry, music or dance). It is signalled and its indeterminacy is defined more precisely by other theatrical signifiers, which in themselves do not have to display a temporal aspect or be an instrument known from a reality beyond the stage for measuring time. As indicated above, time is a relationship between events. It is worth paying attention therefore to the specificity of presented time. For example, for most spectators it is obvious that time is an element uniting the orientation, the cause-and-effect sequence or the teleology of most of the elements, events and actions of the performance, but simultaneously that empirically measured time will almost never be consistent in the tempo of its course with fictional time. Time slips away unnoticed between particular scenes and acts, and even within the ambit of particular utterances by the figures, although it might seem that linguistic articulation in the theatre would have to pass with the same tempo as in the reality beyond the stage. However, in the theatre it does not have to, for stage speech does not have to be a sign of an actual utterance in the fictional realm: it may, for instance, be a sign of mental processes, a state of mind or emotions, which need not be set within any specific flow of time. This is clearly visible during dialogues, which objectively demand a specific time to occur, when related to events taking place in the invisible world offstage. This is always the case when the events taking place in the latter are related by someone who is present on the stage, as in "live" reports of battle scenes (as in *Anthony and Cleopatra*). Also, some scenes, like executions or sexual intercourses are taken off the stage, while other figures remain on the stage engaged in dialogue. The time of these seemingly parallel scenes cannot be treated as belonging to the same stream, for this would lead us to nonsensical conclusions, resulting in suspicion of *eiaculatio precox*.[6]

For many it will not be obvious that the sum of the real time of particular scenes will be different from the sum of the implied time within the same scenes in the fictional realm. It could result from this that, even in a naturalistic convention, within a single scene we might have a different time from the empirical one. Moreover, on closer analysis, it turns out that in each of the scenes time can pass with a different tempo, and this does not refer in the least to the sense of "fast" or "slow" action, but in the sense of the passage of implied time, its condensation, elasticity, in relation to the measurable time of a particular scene. In fact, the tempo of its passage can be different within a single verbal utterance and not even necessarily in a dialogue! Thus the structure of the whole of the performance includes within itself subordinate temporal structures, embracing particular scenes, and these contain structures, subordinate to themselves, including particular threads of the main plot and sub-plots, and also ancillary, non-plot events. Alongside spatial relations, relevant to individual stage pictures, temporal connections have a fundamental significance both in the modelling and encoding and decoding of a performance. The mental time of the spectator, his or her subjective feeling of the passage of time during the reception of a given performance is something else altogether, which is often taken into account by its creators (for example, in order for the spectator to receive the performance as a temporal *continuum*), and thus has an influence on the encoding processes and on the final cognitive shape of the work.

The parallelism of temporal courses does not therefore have to mean an equal tempo of their passing, their synchrony; what is more, it seems that on account of the economy of time, namely that which can and should be shown on stage, dramatists (and/or directors) relegate part of the fabulous events to beyond the stage, in order to "save" time: *diegetic time*, operating there by implication only, becomes conventional to an even greater degree since it is not confronted with the time of the behaviour of the figures visible on the stage (and for these it is difficult in a visible way to indicate gaps in time or temporal short-cuts). It ceases to be a perceptible relationship between events. In accordance with what I said, in each scene this co-efficient time can be different, and what is more, it can – I repeat – acquire various values within one temporal course of a single scene! This means that the scene also has subordinate temporal structures, applying, for example, to events played out in parallel (but not synchronously). That is why the mimed performance of "play-within-a-play" scenes, lasting, say, two minutes of "our" time, becomes a sign of a performance at the royal court, lasting, say, an hour (see Chapter 9). A hunting scene played out in a dance arrangement on the stage will take, say, three minutes, whereas it will be a sign of an all-day-long event. This same hunting could take place off the stage in a short break between scenes – then the saving of time will be even greater. Playing a five-minute-long scene while one of the figures is sleeping does not at all mean that the period of sleep lasted five minutes! From the dialogue that follows it could transpire that it lasted two or three or nine hours. Not understanding the relativity of scenic time in the relation between one event and another or to empirical time leads to interpretational misunderstandings.

The appearance of fictional temporal units explains why within "two hours traffic" we can have a whole life or generation presented on the stage (in the form of "little lives", to use Prospero's wording), a scene lasting five minutes may be equivalent to five hours, and a three-day battle will be shown in three minutes of scenic time. The theatre has the ability to jump "o'er times,/ Turning th'accomplishments of many years/ Into an hour-glass" (Prologue, *Henry V*, 29–31).[7] In *Titus Andronicus* the lascivious Empress Tamora conceives in act 2 and gives birth in act 5, which in performance time means roughly two or three hours. In the same play, in scene 1, the Empress's two sons are taken off the stage for execution and sacrifice, and when their dismembered bodies are brought back to the stage, only thirteen verses of dialogue have passed, which means roughly one minute of scenic time. Years may pass between the acts, and even within a single act, which lasts, say, half an hour of our time, we may be informed that a whole day has passed. The opening scene in *Hamlet* begins at midnight: "'Tis now struck twelve", says Barnardo in line 7. And in line 142 the "cock crows", which means that at least several hours have passed. The dialogue occupying 135 lines does not require more than, say, seven or eight minutes, but it is obvious that much more time has passed within the fictional stream. In *Romeo and Juliet*, the ball at the Capulets' begins at line 14, of act 1, scene 5, and ends at line 144: even if we include dances this cannot take more than ten minutes of scenic time, yet at least four or five hours is supposed to have passed. In this particular example one minute of the "real" staging time is "worth" twenty-five or thirty minutes of the fictional time, which flows within a different stream. But this does not mean that every minute has an equal temporal value. Take the last speech of Faustus in Christopher Marlowe's play, which begins with: "Ah, Faustus,/ Now thou hast but one bare hour to live", and after roughly fifty lines the clock strikes twelve. The soliloquy may be uttered in about three minutes of scenic time, which makes one minute equivalent to about twenty minutes of the fictional time. Even more interesting is the fact that half an hour does not fall in line 25, but in 31, which implies that for Faustus time accelerates (each moment of the second half of the soliloquy is worth "more" fictional time), as is congruous with the psychology of the condemned before their execution. Thus, as we can see, the two present times, although parallel, do not have to flow with the same relation of their pace value.

Segmentation of time

As usual there exist many typologies of time in theatre, but I do not intend to go deeply into the controversies and the discussion of variant theories here.[8] It is sufficient to look at Patrice Pavis's *Dictionary of Theatrical Terms* in order to see how great is the degree of difficulty in grasping the essence of the phenomenon under discussion. I will, however, in order to illustrate the complexity of the problem, differentiate at least a few different times, which, I believe, have significance for the understanding of each performance:

1 Fictional historical time, implied by the performance (this means including also that which occurred before the beginning of the action and that which will occur afterwards, on the condition, obviously, that this is signalled in the scenic text); Pavis calls this the extra-scenic time or dramatic time.

2 The fictional (fabulous) time of the performance, including the time which passes (on the stage and beyond it) from a fictional moment x, in which the action begins (for example tea-time in the salon, Monday, the winter of 1825), until the moment of its end (say, a month later); Pavis does not differentiate this time from the time in point (1).

3 The sum of scenic time, fictional (by which we understand the time of each scene of the performance), which usually will not be equal to the time in (2), since between the scenes there are always gaps in time (only in the case of one-act plays do these times have similar values); Pavis does not differentiate this time either; for him this will still be dramatic time.

4 The fictional scenic time of particular scenes (NB: this will be different not only in the length measured by the watches of the spectators, but perhaps in itself, which means within particular scenes, it demarcates various levels of tempo or condensation of main and sub-threads, plot and non-plot events); this is also a category not mentioned by Pavis.

5 The real time of the duration of particular scenes (i.e. biological time of the actors); Pavis calls this scenic time, although according to him this is a category, which includes the spectator as well.

6 The real time of the duration of the performance as an event (with breaks between scenes, the entr'acte and so on); for Pavis this is also scenic time.

7 The sum of the real time of duration of particular scenes.

8 "Systemic" or structural time – belonging to the dominating semiotic system, which also has a temporal nature (for example, music in opera, or dance and music in ballet, verse in dialogues); as far as I know, this type of time – and its relations with the other ones – has been often ignored in theatre studies.

For Pavis the relation between dramatic time and scenic time has decisive influence. It seems, however, that the following temporal relations might be especially interesting for students of theatre: 2:6, 3:5 and 4:5. Research into the relation of systemic time (8) and the remainder could be particularly interesting – here equivalences could depend not only on subordination but also on inconsistency (for example, rhythmic movement and poetic text).

Others could result from a clash between dramatic time (which means that created exclusively in the text of the play) and theatrical time, for example in productions in which the director rearranges the order of scenes, changes the causality or the logic of events, cuts some scenes or introduces their repetition. Each scene consists of at least one configuration or their sequence. The configuration in turn is a temporal sign and requires a defined time for the performance of all its elements, the relations between them and their semantics.

Accordingly, it seems to be legitimate to differentiate between several features of scenic time. First we may talk of the fictional *fabulous time* (i.e. connected with the "story" that unfolds on the stage) – which by its nature has a "leaping", fragmentary nature, unless it is subordinated to the time of the musical system, where the requirement is compositional continuity: an opera cannot be "shortened" without a significant detriment to the whole, a melody line cannot be segmented. It is to a considerable degree an assumed time, requiring from the spectator retrospective reconstruction; moreover it proceeds with various tempi within even one scene. There is also in the performance *non-fabulous time*, applying to ancillary signs and signals, and also *diegetic time*, which applies to the world not shown on the stage, but is only hinted to or described by the stage figures. Non-fabulous time is fragmentary in its chronology, mainly for the reason that it is not always "necessary" for the development of the plot. This can be – but does not have to be – a stream of time similar to our real one (or rather to its structures fixed in our minds), since there is not the slightest need for signalling that items of furniture or other objects are characterized temporally, unless the time leap also has to be signalled by non-verbal means (then they constitute an active element of the plot event until their informational capacity is absorbed by the recipients). It is necessary to the extent of allowing the recipient to decipher and understand the meanings revealed in the moment of the activation of a given sign.

Taking the perceptive abilities of the spectator into consideration is important in that, firstly, the recipient applies a matrix of external time, which is close to his or her experience (and *deixis*), to the multifarious temporal structures of the performance. Thanks to this the spectator can, at that time and retrospectively, reconstruct and establish relations between these structures. This can also be connected with the feature of our mental activity, which is its ability to create structures: it is thanks to this that our minds give a cohesive structure, the features of a homogeneous *continuum*, to even the most distorted stimuli. Secondly, thanks to various time streams, into which fall all the elements within the space of the stage, one can select a small part of them for purely practical reasons: in order for reception to become possible, and all in spite of the uncommonly rich informational potential and with the semiotic and semantic "density" of everything which is seen or heard. Owing to this, and thanks to the uncommon quality of our minds, which manage to differentiate the intensity of concentration, an information over-load does not threaten us and does not hinder or even prevent perception. I have compared this to the process of osmosis. Fragments of the world presented are highlighted by the creators of the performance (and sometimes by our own selves), like a spotlight in the darkness, and that which in a given moment is not in the foreground becomes neutralized, deactivated, absorbed by the unnoticeable *latent time*, in order to fall once again, if need be, into the stream of fabulous time. This phenomenon can be called *temporal fluctuation*.

Let us pay attention to one more thing: the division of the performance into scenes is based on the segmentation of the whole into temporal segments: each scene is a series of events and actions, developing in one time sequence and in one space. But quite often a different tempo is given to events within particular scenes. The breaking of the sequence of complex signs within a scene (for example, by a blackout, by freezing all the figures except for one or by having all the figures leave the stage) is an important signal for the recipient, since it implies a passage to the next scene, and therefore a gap in time or a partial – because usually it applies to just one element of the scenic text – suspension of the fabulous time. This is intended to emphasize this particular element of the performance, not necessarily a verbal one, in order to allow it to function to a certain extent in two times, within a given temporal course of action, but also outside a given time sequence. This temporal duality will, however, almost always be legible only for the spectators, not for the figures, who notice nothing. It must be added that in this very way are created the allusiveness and referentiality of the scenic text, and also its ability to differentiate one selected element: remaining an element of a given temporal sequence, for a moment it breaks out of it and indicates the possibility of another reading, ascribing it to another time and space, external to what occurs on the stage, e.g. to specific art works, cultural, social or political codes, or to another series of events (for example, political ones), which are following their own course sometimes in the extra-theatrical reality. This is a form of *temporal ostension*, which signals the possibility of a simultaneous reading of the same signifier from two or more temporal perspectives, which inevitably generates several layers of meaning. Thus, theatrical costume might be a signifier of an ancient soldier's outwear (e.g. in *Julius Caesar*); at the same time it may iconographically allude to a much later, say, Baroque painting showing a cunning politician and a tyrant; simultaneously it may allude to a living ruler and his habit of identifying himself with ancient heroes. It may also allude to a film, or a TV programme, or a comic script. In fact, the whole production may be modelled in such a way that it will signal the different temporal perspectives that may be employed to its understanding. It does not have to mean that the classical play is predominantly about a modern ruler, but, usually, it implies that, among other readings, it may be read as such.

Time is the message

It is worth paying attention to the techniques of creating temporal structures in a theatrical performance. Obviously every performance is played out
before our eyes and the time needed for its realization can be measured on
our watches. We are, however, more concerned with fictional, scenic time.
The simplest way of signalling its passage is through references to time arising from the dialogue, and through the mere presence of a live actor. As I
said, the actor is a biological clock set on the stage, a perfect instrument to
signal the present time and the natural flow of time. Thus it is the actor who
is capable of showing time as a continuum. Objects, on the other hand, may
be employed to show the temporal setting or jumps in time, but not the
fictional time as a flowing continuum that has the ability to mark its different tempo or signal jumps in time.[9] There is in fact a whole range of non-
verbal methods (considerably different from those in film) for doing that.
For example the scenography (or a costume) can in each act be different and
emphasize the passage of time, either through ever greater "destruction" or
by the changing fashion of the interior decoration (or fashion of costume) or
the changing portraits of the heads of state. Scenographically one could show
the changing phases of the moon or the seasons of the year. One can tear
pages from a calendar, or place a clock on the set, although this is rather risky,
if the clock is to work properly. A clock showing the "real" time is a highly
undesirable phenomenon in the theatre. It is understandable, because it may
appear in an indexical relationship with an actor, who is constantly signalling a flow of fictional time that is not congruous with the dial (as in Faustus's
final soliloquy). This may cause serious discrepancies the directors usually try
to avoid. For this reason directors usually do not like to have real working
clocks on the stage. In Eugene Ionesco's *Bald Ballerina*, the clock on stage
strikes seventeen times, then seven, then three; then it does not strike at all,
to begin again, in part two, with five strikes, then two and one; towards the
end it strikes like mad. It may be added that there is no relationship between
the time of the clock and the stage action.[10] Similarly, the "crooked" chronology may be found in medieval drama, or in regional varieties of theatre, as in
the Turkish *Ta'zije* theatre, where the cause may appear after the effect.[11]

The passage of time can be signalled in yet other ways. These could be,
for example, gradually introduced technical innovations, constituting the
fixtures and fittings of a flat or kitchen. Various gadgets, changing over time,
the media (for example newspapers) and pieces of apparatus may serve as
time markers. Similarly the actor can in every scene or act create an ever-
older figure. This will be signalled not only by make-up, way of moving,
voice, wig, and so on, but also by costume. This also creates the possibility
of creating various additional meanings, arising from the relations between
temporal structures signalled by the text expressed on stage and those, which
are created by non-verbal signs. For example, a dialogue could be "timeless",

which means that it will not signal the passage of time (the physical time required for the speaking of a defined number of words does not have to be an element of fictional time), whereas the elements of the scenography will emphasize the passage of time, even a considerable one. From the clash of these different temporal structures there could arise metaphorical interpretations, but these could also have an ironic significance. And conversely, in the dialogues the motif of passing and of time will be constantly repeated (as for example in Chekhov), while the space around will not display any changes. Also music can show the passage of time: in each scene the music could be stylistically closer to our times; it could also become a repeated sound effect, indicating simply "a marked passage of time" (like the turning wheels of a speeding locomotive in old films). In any case the temporal structures in the theatre as a rule fulfil the important function of a binder in building the matrix around which the action is played out: in order to understand it, knowledge is necessary about what is being played when, and what is the sequence of events. And, as observed above, time is a relationship between the events. Obviously, in "experimental" or avant-garde theatre we will have attempts to break even this convention down (rather our habit of turning events and history into fables or narratives). There are also theatres (and cultures), such as dance theatre, where temporal chronology, regarded by us as correct, does not have great significance.

A great range of techniques are known which allow for a precise definition of time and its passing, and also tempo, which will be particularly important and noticeable when it is different from the pace of empirical time, or in the case where several temporal structures exist within one scene. One can even try – as was popular in the 1960s (and later) – to reverse the passage of time: each successive scene is played out chronologically earlier. Obviously, within particular scenes this cannot be done, since the figures would not only have to walk backwards, like in a film played in reverse mode, but also speak backwards, which is a physical impossibility (unless it is a recording). There are various intermediary techniques applied in the theatre. For example action is shown, which in the empirical world demands a defined time – say, the roasting of a chicken, which lasts at least an hour. If the roasting of a chicken on the stage lasts five minutes, or even ten, we surmise that time has been condensed and the few scenic minutes are in fact an hour (this, however, undermines the realism and verisimilitude of this scene, particularly in productions with a mimetic orientation). If during this time the housewife is pottering around the kitchen and from time to time says something to her husband, who is engrossed in a newspaper and mutters something unwillingly in return, we understand that this stereotypical scene from family life is lasting at least an hour. Its time is delimited by the process of roasting the chicken. Thus a culturally defined extra-scenic time will clarify and define the fictional time on the stage (not the other way round!), demarcate its tempo and frame, all on condition that the temporal structures of the elements of the action, those shown on the stage and those which we know at first hand, will be contrasted with each other.

Sometimes, in order to avoid this contrast, directors take various measures: for example, the scene does not begin with putting the uncooked chicken into the oven, but it will already be roasting and only the end of the process will be shown. A great deal of time is necessary in order to dig a grave – that is why Shakespeare began Act V of *Hamlet* from the scene with the grave-diggers at the moment when the grave is almost dug (the dialogues and the songs allow the digging to be completed; one has to remember, however, that in Elizabethan theatres there was no act division). The actual ceremony of the funeral is also a noticeable abbreviation. Obviously, it is difficult on stage to speed up the making of toast, the boiling of an egg or morning shaving or showering without undermining the credibility of these scenes. You cannot cut out or edit the scene of getting dressed or going to the door to open it for guests – the stage figure has to hear the bell or knocking, get up from the armchair, put aside the newspaper, approach the door, open it, allow the guests in, accompany them to the interior and so on. In similar, particularly "short" scenes in the theatre, no temporal short cuts are possible (you cannot shorten the time of approaching the window or putting on a coat or opening an envelope). But these activities can have a different time value ascribed to them than the one which they would have in the extra-scenic reality. For example, by confrontation with a parallel – but not synchronous – thread of the scenic action, or event, which will have a defined temporal value. Then – in an attempt to bring to order the developing temporal chaos – we transfer the temporal value from one level to the other and accept that during the dialogue, which was defined (for example verbally) as being three minutes long, an egg was boiled (although according to our watches the dialogue lasted a minute!) or – if it was not defined temporally – we impose on it the temporal dimension of boiling an egg, which we know from experience. On a similar principle we can also show longer scenes conventionally by a considerable shortening of the sequence of compound signs (for example a battle, duel or fight can be shown with the help of a choreographic arrangement). In some types of theatre art (Beijing opera, Kabuki) we have numerous conventional and schematized scenes, repeated in various performances, whose meaning does not change and is recognized by the experienced spectator. That which is defined in a given fragment of the performance in its temporal dimension imposes that dimension on the parallel series of events or actions, even if they are – measured by extra-theatrical physical time – not synchronous. Hence the next conclusion: that in reception we will always try to simplify the quantity and variety of temporal structures in order to unite or blend them in one *continuum*. In the opposite case, there is the threat of temporal chaos.

The contrast and confrontation of scenes played out in a temporal abbreviation with those in which time has a tempo close to physical time can be a conscious and significant measure taken by a director. This can clearly be seen in the work of Eimuntas Nekrošius, for example in his staging of *Othello* in 2001.

There is a scene in the production, in which, after murdering Desdemona, the eponymous hero slowly begins to realize the horror of his act. Around the body of his victim he arranges flowers, sits, rises, looks at the composition he has created, improves the arrangement of the flowers, sits again, freezes and so on for a further five or maybe more minutes. Apparently nothing is happening in the sense of the plot action, but let us pay attention to the fact that scenic time has become equal with the physical time of the spectators, through which the pain of the figure becomes even more moving, and the merging of the two times creates an uncommon tension (the spectator begins to share time with the figure). Since, however, time in the previous scenes had been conventional and condensed, the present "merging" of times is only apparent. The spectator has become accustomed to the fact that each minute of scenic time has a different value in the physical world – it is "longer". That is why, in contrast with what precedes the scene under discussion, its time acquires the feature of an even greater extension, making it last even longer and, like Othello's pain, it becomes unbearable. This is at the same time yet another example of the possibility of imposing on the spectator a feeling of participation – through the community (with the figure) of subjectively experiencing time. Time becomes the message.

We see therefore that time forms quite complicated structures within particular scenes, since in many plays and productions, temporal streams can proceed in parallel or they can branch off in order to join again some time later, many times tying together various strands of one plot or more. They can be simultaneous (synchronous) or move with different tempi. In the case of several plot threads at the end of a play, we will as a rule be witnesses of a joining of their times and spaces: the simultaneous appearance of all the figures enables misunderstandings to be cleared up and liquidates all discords and discrepancies in the levels of awareness of particular figures, and of the spectators. Now the spectators will be able in retrospect to reconstruct the development of events, join them in a homogeneous temporal structure, which until then was not clear to them, mostly because – for example – a considerable part of the events was relegated to off stage. They are therefore dynamic structures, and, what is more, those which we do not see on stage have the significant function of supporting the plot action, enabling scenic time to be saved, adding credibility to the development of events and also constituting an important instrument for manipulating the awareness of the recipient.

Let us also notice that scenic time is strictly connected with scenic space and always emphasizes the differences dividing it from real space. On a similar principle the essence of scenic time is often different from the time measured by our watches, once flowing slowly, then again quickly. This is a phenomenon determined not only by psychological factors but also culturally. But it is we who give time its pace and its subjectivization is the result of our thought processes or our awareness. In the theatre – in contrast – we do not have any influence over the tempo of a given scenic time, since it is preordained and is in fact a component of a projection of someone's view of the world and someone's thought processes or awareness (the fact that during a performance time may seem to be dragging or rushing by unnoticed does not apply to fictional scenic time, but to real time). That is why it should be remembered that the subjectivization and relativization of scenic time is independent of our thought processes or our perception of the world; it is given to us in such a dimension as the creators of the performance wish. Let us also remember that the scenic space is oriented and thus is historical, being played out (modelled) in a clearly defined and pre-ordained time. All this empowers us to state that within particular scenes we can be dealing with various temporal structures of various events and this does not at all have to lead to temporal chaos.

It can therefore be said that in essence we are dealing in each performance with several streams of time, of which in traditional dramatic theatre the most significant is the one constituting the temporal matrix of the plot. The creators of the performance construct it, creating scenes and events whose time and tempo are also varied (as we see, changes of this tempo can be significant). In order to construct a defined scenic event or action, the appropriate range of verbal and non-verbal signs is chosen, and they are linked not only spatially but also temporally: in this moment, say, the music, text, gesture, property and mimicry "flow", playing in the same temporal stream, then after a while they create another configuration, with different constituent elements. In other words, various configurations of signifiers are pulled into or fall out of temporal sequences, which constitute the matrix of plot threads, or also parallel secondary threads, and even – through connotation – become an element of events or plots playing out (in their temporal sequences) outside the theatrical world (for example, political events or social conflicts). On this also depends the sign mobility of all material substances constituting the performance, and also their ability to constantly transform themselves into various types of signs. That is why we should not consider temporal relations in the theatre as homogeneous and applicable exclusively to the scenic action or plot. It seems that the decisive factor here is the domination of one of the semiotic systems on the syntagmatic axis. In the case of the domination of temporal signs, temporal structures will be vital for an understanding of the whole; in the case of the dominance of spatial signs (for example, the scenography) simultaneous perception becomes, at least to some extent, possible, and consequently with the partial omission of temporal structures or with a significant lessening of their role. This was the case, for instance, of the Stuart masque. There can also exist performances, where we will have an alternating domination of various types of signs.

Reading time

On several occasions I have mentioned the spatio–temporal division of the stage and the auditorium, which is necessary for the creation of fiction in the theatre. Let us remember, however, that it is either time or space, if these two can be separated at all, that is sufficient for such a distance to be created. The grounding of this division only on time is, however, less common in the theatre and that is why it requires at least a minimal illustration. In John Marston's play *The Malcontent* from about 1604 we have a wonderful "induction" scene (written by John Webster), in which there appear actors (and two "spectators"), who create the figures representing themselves! The scene shows them during the preparation for this very performance; it is full of banter, jokes, caustic remarks and allusions to theatrical events in London. For us, however, what is important is that the figures carry the names of the actors who are playing them. So – despite what I said earlier – can an actor play him/herself? He can, on condition that there is a shift in time: let us notice that the figures are playing a scene from before the beginning of the performance, they are still in the tiring-house, and not on the stage. There is still, say an hour until the performance, and there is also a spatial shift (the tiring-house of course is not on the stage). Thus an actor can play him/herself, but he/she will only be him/herself in some more or less distant historical time. Without this shift the afore-mentioned scene would present an absurd situation, in which the actors had a dressing room (tiring-house) on the stage and were improvising/gossiping before the spectators, and all this at a time when they should already be acting. Furthermore they do not notice the spectators, so if this is not yet the performance, this means not so much that the "fourth wall" has fallen down, but that the actors are making light of the spectators, and are quite simply ignoring their presence. In the latter case there will be no fictional time and space and therefore no theatrical performance. Although such an artistic idea is conceivable as temporarily taking place on the stage, it in a drastic way destroys the possibility of the creation of theatricality. So, in this particular scene the rule of accessibility is fully observed, but the sitance between the two spheres is marked mainly by the dual time. Let us notice that in such cases the time of the "introduction" flows more quickly than the real time of the auditorium implied in the text, thanks to which the temporal distance with each moment is becoming smaller, to disappear entirely at the moment of the fictional beginning of the "real" performance. In the opposite case the performance would never come about, since the events portrayed on the stage would always be happening "earlier". Now actors change their deictic costumes to become fictional figures from Marston's play. A similar trick occurs in Molière's one-act play *L'Impromptu de Versailles* (written in 1663), where both the actor-director and his actors appear under their own names. We see the rehearsal before the premiere, and it turns out that the actors do not know their lines. They are saved from embarrassment at the last minute by the decision of the king to postpone the premiere. Here, however, we also have a temporal shift: the actors are shown before the premiere, so, as we can see them, it is not during the performance but during a rehearsal. Thanks to the condensation (and acceleration) of the scenic time it eventually catches up with the time of the spectators, so that, at the appropriate moment, just before the curtain is raised, we can see a happy ending.

The relations of the main and sub-plot threads can be more complicated, but they always depend on a temporal contrast. In the performance of *Hamlet* by the Theatre of Farce from St. Petersburg (1999/2000), directed by Victor Kramer,[12] from the beginning there appears a mysterious figure of a poor man (we can tell this from his costume), who comes to an inn (he defines this space himself, by his presence and gestures), sits at a table, eats and drinks (the tables in this performance fulfil various functions – they are even the stream in which Ophelia drowns herself; this is testimony to their considerable sign mobility). He says nothing and we do not know who he is or what he signifies, particularly as those spectators who know the content of the play know that such a figure does not exist in the Shakespearean original. From the beginning he evokes surprise and curiosity: in the end we want to know who on earth he is. But the director for a long time gives us no answer (this technique is called retardation). What is more, the space of the "Stranger" is clearly separated from the space of the plot threads, and also takes place in its own time, which means that it can be considered as a space in which there develops a – mimed – sub-plot thread. But that does not occur in parallel. That is, it is not synchronized with the main plot, its time flows with a different pace. Obviously the figures from the main plot do not notice the presence of the "Stranger" (and this is reciprocated: he does not notice them), which emphasizes the temporal and spatial distance between them. This is at the same time an example of a mutual suspension of the rules of accessibility: in contrast to a theatrical performance, neither side can look into the reality of the other side – hence the conclusion that they are divided either by time or by space, or by both simultaneously. It is only towards the end of the play, namely in act 5 of Shakespeare's original, that we find out that the "Stranger" is in fact the Grave-digger – the recognition occurs at the moment when his time and space are joined with the time and space of the plot of the performance.

Retrospectively we begin to understand that on the one hand we have been shown a man who, taking advantage of a break from work, has gone to an inn to have something to eat; when he has eaten and drunk his fill, he returns to the cemetery where he encounters – spatio-temporally, it could be said – the plot. In this sense his time at the beginning of the play considerably foreruns (by several months!) the fabulous time, since if we consider them together it would be only, say, a couple of hours directly before the scene of digging Ophelia's grave (when the time of the plot catches up with him). In a word, the time of the Grave-digger runs with a different tempo from that of the fabulous time of the play. This is of course a "realistic" interpretation. We have here, however, the possibility of another interpretation, since we can notice the temporal contradiction, because on the stage two times are being shown in parallel, but not simultaneously (until the entrance of the Grave-digger into the mainstream of fabulous time). We begin to understand that the fabulous time of the "Stranger" has become somewhat extended: that which normally lasts two hours here lasts a few months, which is a (culinary) improbability: we therefore start to guess that the time of the Grave-digger has apparently been extended in order for it (as a sign) to signify something different – quite simply, for him to be constantly present on the stage (his function is permanent presence). In reality the time of the Grave-digger flows with a different tempo, close maybe to the time shown on the watches of the spectators, but not necessarily. In retrospect we see that on the one hand this could mean "routine": each of the Grave-digger's days looks the same, so that even the extension of one day into several months can be justified, on the other hand he becomes a legible symbolic sign, or perhaps a metaphorical one. All the actions of the figures on the stage are accompanied by the Grave-digger, whose presence they do not notice (the first grave-digger was Adam – as Shakespeare reminds us through his words – a sign of inevitable death). The fate of a human being is from the very beginning connected with this profession, the metaphor of which becomes the permanent presence of the Grave-digger on the stage. His meaning is clarified by the specific temporal contrast between two threads of the plot – the main and the sub-plot. The full meaning (or maybe meanings) of this contrast will only be revealed at the moment of their connection. Then the two times begin to flow in one stream, in one direction and with one tempo. We see therefore once again what great significance in the art of the theatre is due to the spatio-temporal division. The parallel layering of non-synchronous temporal structures can lead to metaphorical decipherings.

The afore-mentioned diegetic time in turn applies to everything which is not directly shown on the stage – as a rule this is a scene which the playwright or director has "ejected" from our sight and which is played out off stage. This signals its beginning, next its absence, and afterwards a return to the plot played out on the stage. Often implied historical time is diegetic as it includes the events forming the temporal frame of the "macro" plot thread, and also various sub-plots. In *Hamlet* it includes the events from the murder of the old king to the assumption of power by Fortinbras, therefore a period of several (six?) months. This is therefore linear and continuous time, creating a usually passive background or frame for the fabulous time, consisting of irregularly connected fragments. It is also time, which is "lying in wait", only to become activated at the appropriate moment. Sometimes it merges with the fabulous time and then it becomes activated, bringing its own, hitherto unseen, fable, which as a rule subordinates to itself what is being played out on the stage, destroying it or absorbing it (like, for example, the Great Mechanism of History, introduced to the theatre by Jan Kott). An example of this could be the previously mentioned intrusion of Fortinbras (and his time), or that of Laertes (after his return from France), into the world of *Hamlet*, hitherto being played out in the space outside of the castle of Elsinore. A private history becomes absorbed by the time of history. Quite often the plot thread shown on the stage is led to the moment of its temporal merging with diegetic time, that is, the implied time of history, flowing independently of that which is shown on the stage. This is emphasized by the connection of both temporal courses and their mutual interaction: that which is shown on the stage can have an influence on history or vice versa – it is history, which brings an end or a solution to the plot being shown.

There is yet another temporal aspect in the Kramer performance, connected with movement and motion. Cognitive psychology has proven that there is a close connection between motion and an individual experience of time.[13] There can be no doubt, that generally in theatre it is hard for the creators to achieve the intended shared experience of time of the audience with the stage figures, a sensation which may result in the illusion of being somewhere else at a different time; but it is not impossible. Let me recall here that the missing link in the illusionistic participation in the world of fiction seems to be – apart from spatio-temporal synchronicity – movement. Falling under illusion is one of the symptoms of theatre perception. But on certain occasions it is the shared movement of actors and spectators, concomitant to the shared sensation of time, that brings the physics of the different realms very close together. We must not forget that the phenomenon of perceiving space is completely different in the theatre than in our every-day life experience. There can be no doubt that we perceive space fully thanks to the possibility of movement and the changing point of view. In theatre, space is created and shaped by man for our viewing; it is there to be "read". It is oriented and may reveal a temporal value. The actors move freely within the space given, and signal us the way in which it is perceived by the fictional figures. Usually, when seated in the auditorium, we do not and cannot share the movement with the actors. If this were possible, we would also experience common time (since movement within a shared space requires the same time).

Let us repeat the question, therefore: is this at all possible? And let us immediately answer that it is. In Victor Kramer's production of *Hamlet*, the effect of the "movement" of the spectator was brought about by placing the auditorium centrally, on a revolving stage (the spectators at the festival *Teatr Muzyczny* in Gdynia, Poland, were seated on rising benches). The action took place around the revolve, on the stationary part of the stage. When it was necessary to change the space, the *ad hoc* auditorium revolved, which brought about an uncommon situation of reception: the auditorium was in motion and followed the action from one space to another. A 180-degree turn meant that what until then had been invisible behind the audience's backs now became a new "scene". This did not, however, cause a break in the action. There was no break for the change of space, for getting there – the actors continued to act, they moved in the direction of the revolution, still remaining on the stationary part of the stage. In this way we moved along with the actors from one space to another. This does not happen in theatre very often. This also meant that at times the tempo of the figures' steps was identical with the tempo of the revolution of the stage and of the spectators sitting on it. Now it was we, the spectators, who moved towards a new space, and some of the actors (e.g. Hamlet and Horatio on their way to the cemetery) followed along. The effect was quite uncommon, since in this type of arrangement of space changes there was synchronicity of movement and its tempi of the auditorium and the actors (and therefore of time); in consequence there was a momentary feeling of common participation in the events portrayed: we were not only passive spectators, but also "virtual witnesses" of the events. In this way, movement meant that the spectators took over – at least for a certain time – the point of view (vector of viewing) and the time of the figures. It may even be said that an illusionary and momentary indexical relationship was established between the spectator and the fictional figure. And therefore the scenic space ceased to be a projection of someone's view of the world, i.e. a space oriented and modelled by someone for display, and became for a short while a real space perceived from our individual and changing point of view. As is always the case, we – through our movement, movement through time and space – gave it shape. And apparently this was enough to abolish or suspend – at least for a while – the permanent hiatus between the view of the world, in its spatial and temporal aspect, of the figures and that noticed by the recipients, between the oriented fictional timespace and the physical timespace. For succumbing to illusion depends on the acceptance by the spectator of the point of view of the fictional figures and of their way of perceiving the world. When the two models of perception merge, the split present time becomes one.[14] Thus, the experience of fictional time becomes real, and not necessarily just an intellectual construct.

Time-generating components

Since time and space are the most important constructive elements of theatre, we may divide the related phenomenal components into two categories: those that have the ability on their own to generate temporal structures, such as human bodies, stage speech, music, dance and the like, and those that do not reveal that ability, such as objects, silence and light, but may gain the function of a time and space marker through the course of the action.[15] Of the first category we may distinguish subcategories: one, including those components that generate either time or space, and, two, those that are employed in both time and space construction. The first of these may be labelled as TGC (Time-Generating Component) or SGC (Space-Generating Component), whereas the second subcategory may be labelled as TSGC (Time- and Space-Generating Component). Of course, the same or similar material substance of theatre signs may perform different functions, depending on the circumstances. For instance, the sound of a radio may function as a TGC when there is no shift in space between the scenes, and it may also function as a SGC, when the shift occurs. Light may indicate the time of a scene, and therefore it may signal, through, say, the change of colour, jumps in time, forward or backward, between the scenes, but it may also indicate jumps in space and will therefore function as a SGC. In both of these instances, of course, we might have situations when the sound of the radio and light indicate a jump in both time and space, and then they will function as TSGCs.

The functions listed, especially those belonging to the second category, i.e. signs that are not capable of signalling time and space on their own, need an additional factor that will enable the spectator to read them correctly, and that factor is provided by the stage fictional figures who continually signal to us – through the actors – the way they perceive the world around them, i.e. the fictional realm, and that mode of perception contradicts that what we see and hear on the stage. Thus, within the same space, say, a room in somebody's house, jumps in time may occur, and these may be signalled by an acoustic TGC; for instance, a distant sound of cannons being fired. From the ensuing expository dialogue, supported or not by the stage set, we understand that the action is taking place during World War II, in some German town, and the Russian front is approaching. In turn, the same space without the sounds of cannons may indicate the time before or after the war, depending on how the stage figures define it. Once we learn the temporal relationship between the sounds off stage, or the lack of sounds, and the action, we do not need the stage figures to define the time for us any longer: the information is conveyed by the sound itself or by silence. Thus, the offstage sounds alone will have the ability to signal a jump in time (or space or both), backwards or forward. This explains the theatre's ability to endow non-semantic components, such as light or music, with significance.

Among TSGCs, a special case is represented by those that reveal the ability to signal on their own the temporal structures employed. This ability derives from the fact that TGCs are temporal signs, i.e. they require a certain amount of time to reveal their entire selves. Music may stand as a good example here. Of course, we have to distinguish between the music that is heard by the stage figures and the music that is not heard by them. In the first instance, a temporal and spatial contiguity is created, by which an indexical relationship is established between the music and the fictional figures. In the other case, the music is not part of the fictional realm, and has to be treated as an external, hence metatheatrical, addition made by someone who is responsible for the production. In the latter case, the music does not have to indicate any specific time and space, and the same tune may be sounded for a changed scene, taking place in a different space and at a different time (as is often the case in films). However, when the music is perceived by the figures as music, it becomes a very conspicuous time and space marker, without the need to define it further verbally. Thus, it is a TSGC. Moreover, it may be noted that whenever music is perceived by the stage figures, the time of the music and the fictional time of the figures merge, and in most cases they flow with the same tempo as the time of the spectators. I write "in most cases", because in non-mimetic productions music may slow down or accelerate, or may be attributed a different temporal value, which inevitably creates a contrast with the time of the spectators, which does not and cannot coalesce with the awkward acoustic phenomena on the stage. And one more thing: TGCs have the ability not only to become markers of specific time, but also to transfer temporal values to other objects and phenomena within a given space, i.e. to those components which independently do not have that ability. This is achieved through the establishment of an indexical relationship, which rests upon the spatial contiguity and/or upon the merger into one cause and effect sequence. This explains why in theatre an apple touched by the biblical Eve becomes a fruit in Paradise, whereas the same apple touched by a modern woman jumps in time to become a present day object. Moreover, even if the modern woman is not named Eve, a certain equivalence is created, by which some of Eve's attributes are transferred through the time-faring apple to the contemporary person. This in fact shows us how meaning is created in theatre and how it is conveyed from one sign to another. Thus we see the communication vessels at work, and time seems to play a decisive role here.

All of these factors have to be taken into account when discussing plays that employ non-verbal TGCs. In the very last scene of Tom Stoppard's *Arcadia*, a play that lends itself perfectly to discussion on practically all theoretical issues connected with theatre and drama, the situation becomes highly complex: there are two plots from two distinct time streams that spatially merge and are presented simultaneously.[16] Up to this scene, the plots are presented alternately. This, indeed, is a scene that is among the best ever created in drama. The whole play takes place in one space (the hall of an English aristocrat's house) and rests on two related plots set on two different time levels: the beginning of the nineteenth century and the end of the twentieth. We have two groups of figures belonging to two different historical times. The thread of the plot, meshing, applying to a considerable degree to both time levels, is played out in an alternating sequence of scenes: one from the beginning of the nineteenth century, one from the end of the twentieth, and so on. All the items of furniture, and even most of the properties, various bibelots, remain the same (though sometimes their function changes). In the last scene, however, Stoppard combines all the figures spatially, which means that there takes place a mixing of historical figures with contemporary ones and there is a simultaneous development of the two plot threads. Obviously, this has to be carefully directed so that the figures in their invisible time "veils" do not bump into each other and mutually disturb one another. This demands a careful distribution of activities and utterances between the two groups of figures: for most of the time they cannot be active simultaneously, and the spectator has to know which of the signals emanating from the stage are to be treated as related spatially and/or temporally. This is why, as in an orchestra, some of the stage instruments have to be "silenced", while the others sound to create the desired acoustic effect.

We are dealing here, therefore, with a layering of two temporal levels and it is time, which constitutes the factor separating the figures. The space remains as if the same, but the space from the beginning of the nineteenth century is imposed on that from the twentieth (or rather the other way round). But for the spectator in the theatre, the layering of properties and the physical space is not noticeable. Stoppard's game with conventions does not end here, however, for in the same scene some figures from both time levels sit next to each other around a table and even read from the same sources (Thomasina's exercise-book) or drink wine from the same glasses! What in this case happens with the wineglass? Or with the wine which fills it? At the moment when it is being touched by a person from the beginning of the nineteenth century (being a TGC), the wine-glass becomes activated, torn from its time (or rather a-temporality) and, having formed with the hand holding it (and thus with the figure, who becomes the link with the space) the composite material of a compound sign, joined in a new configuration which signals a change of indexical relation, it becomes the sign of a wine-glass from the beginning of the nineteenth century. It is this touch which causes the object to acquire, together with the hand, the indexical relation and hence the quality of a sign to be grounded in a specific time. Set aside on the table, it is deprived of spatial contiguity with the actor and its signifying (and temporal) value significantly decreases. Objects by themselves cannot create a stream of fictional time. It is only live actors who can do this, because they are the most important TGCs in theatre. Also, other phenomena that are produced by humans, such as singing, dancing, radio or television programmes, or music making have the ability to denote the fictional present time. Consequently, in Stoppard's play, when the wine-glass is picked up by the hand of an actor impersonating a figure from the end of the twentieth century (another TGC), the same wine-glass (and the wine inside) undergoes another metamorphosis and – joined in a new configuration and signalling a new indexical relation – becomes transformed into the sign of a contemporary wine-glass and wine. We guess also that the wine even changes in taste, although we cannot verify this empirically. In this way the whole interior is divided temporally: it is simultaneously an interior from the nineteenth and the twentieth centuries, the one superimposed on the other. It would be hard to find a better example of the magic of the theatre. Let us add that the layering on one another of configurations of compound signs means that the role of those elements which enable them to be differentiated, increases. In this case, for a brief while it is the wine-glass, which becomes the dominant sign, transporting, like a time machine, back and forth, all the other objects from one epoch to another (although, strictly speaking, this is done through the changing indexical relationships). Since this is noticeable for the spectators, the playing with conventions in particular stagings of *Arcadia* is a source of humour. Two fictional semiotic orders are shown and juxtaposed, and the humour stems from the temporal paradox, for the same material substance of the production is shown as belonging to two different time streams, which overlap and appear in one sequence. The theatre presents its five dimensions in action.

6
Sculpting the Language, or Stage Speech

Theatricality of language

From the discussion above it appears that even the most complicated theatre phenomena may be explained and elucidated by analysis of the time structures involved. When language is the object of scrutiny, however, the situation becomes particularly complex.[1] The raw material of natural language (speech) is air, or more precisely, sound waves of a particular frequency, emitted by the vocal apparatus. In describing the chemical composition of air and presenting diagrams of wave frequencies in the process of verbal articulation, we do not even touch upon the complicated problem of language as a sign system, particularly in such uncommon uses of language as we find in artistic texts or when it is one of the constitutive substances in a performance. In theatre, what we hear is the articulation of the factual speech of the actors, not of the figures. Language becomes a sign of a sign, even more than that. The utterances of the figures are only implied, they are not audible. Thus, the initial premise that I shall attempt to prove in this chapter is that the language used by the actors is transformed into an iconic system of signs, and above all becomes a sign of the natural language used by fictional figures.[2] What follows is that the meaning of signs that involve verbal utterances is generated by the juxtaposition of the direct meanings of the latter within the phenomenal stage (and the real world outside, which naturally includes other cultural texts and codes) and the denoted, hence indirect, meanings within the fictional realm. Moreover, the stage language is accompanied by all sorts of sounds, movements, mimicry, gestures provided by the actors and also by the visual material context. All of these constitute the bundles of signals that simulate the brain and the body of the recipient, activating his/her own experience and understanding. This means that it is not the language alone that "performs" the function of activating the spectators' cognitive processes. According to recent research in neuroscience, the language and the behaviour of the actor make the spectators experience and feel parallel states rather than just perceive information about someone else's feelings or emotions. It seems that in the theatre the ability of the language to activate the spectators' own experience in enhanced by scenic tautology, as described in the earlier in this book (Chapter 2). As Amy Cook has put it, "language is less a system of *communicating* experience than actually *being* experience; we do not translate words into perceptions, we perceive in order to understand".[3]

Traditionally, in literary studies the language in theatre is treated as a form of existence of literature, i.e. a scenic articulation of a literary work, dressed, perhaps, in theatrical costume. However, that costume is essential here, for as is the case with all the other systems of signification employed on the stage, language likewise, in the process of transmutation, loses some of its intrinsic systemic features and becomes something else altogether. This "something else" demands a deeper scrutiny, but let me start with a simple observation: it seems that "language" is not the right word to use in reference to what is actually articulated on the stage. Many features revealed by the peculiar language used on the stage empower us to perceive drama as one of several systems transmuted into a new medium and thus constituting the performance, within which it will first appear not as a text set among other "texts" but rather – blended with them, through which it partly loses its independence as a literary text and becomes subordinated to the overall structure of the performance which, in turn, is not entirely linguistic. Thus, paradoxically, words lose at least some of their verbal qualities, and become, as Eli Rozik has rightly observed, basically iconic signs.[4] This is the immediate consequence of the appearance of the theatricality effect, resulting in the rise of the dual function of every scenic component. And the effect includes the language, which becomes blended with the material context of the stage, and simultaneously refers to two realms, the fictional and the phenomenological.

However, we have to distinguish between at least two types of the relationship between the language and its immediate material context of the scene. If we take the Elizabethan stage as an example, we shall notice that the same stage in its unchanged physical appearance is used for different productions, which means that the relationship between the language and the phenomenology of the stage is not unique for a given show, and is basically functional and open to the changing symbolic meanings (as discussed elsewhere in this book). The material context for sundry productions remains virtually the same. But the meaning of this context is open to constant changes, due to its ability to absorb meanings from acting. The stage becomes a semantic chameleon. Thus what decides on the generation of meaning are predominantly the words and the actors' bodies and their various configurations. The situation is totally different if the production is integrated with its immediate material context, and creates meanings generated by all sorts of relationships between the words, acting, changing lights, the set, human bodies and the surrounding signalling matter, incorporated in the performance. This means that the production cannot be presented within a different material context without a significant loss or change of its semantics and aesthetics. Also, this implies, that a different production cannot be presented within this particular space, which had been created *ad hoc* for a particular production, and not any other. It can, of course, but will not make any use of the surrounding context, or, even worse, is likely to produce unwanted connotations. Thus, integrated productions incorporate the stage and its phenomenology into themselves, creating inseparable blends of greater semiotic complexity.

On the whole, following the two modes of spatial orientation discussed in Chapter 4, we may divide theatre productions into two categories, isolated – where the materiality of the productions rests entirely on the language and human bodies of the performers (including costumes, wigs, makeup, props, music, etc.), and integrated – where the materiality of the production includes the phenomenology of the space of the stage. In the latter case, it becomes site specific. On rare occasions, it may even include the space of the theatre (as in liturgical drama or in court theatre), or an area adjacent to it, as was the case of Jan Klata's *Hamlet* (entitled *H.*), staged in 2003 in the Gdańsk shipyard renowned for the Solidarity movement. In the first category, the language and the human body are likely to dominate, while the second one opens the possibility of equal importance, if not the domination, of scenography or any site-specific location of the performance. The relationship between the two is likely to define the style or "genre" of a given show. In today's theatre practice, at least in Europe, the second category plays a leading role, as the term "postdramatic" suggests. Also, it should be noted, the transformation of one category into another brings about not only technical problems, but inevitably changes the meaning.

Thus, in the case of integrated productions, through a magical chemical reaction, following the unique formula of the particular production, language becomes an amalgam with the world of objects, human bodies, light and non-verbal sounds on the stage. It absorbs meaning from objects, and through reciprocation bestows meaning upon the hitherto non-semantic elements (such as light or music). It clarifies the indeterminate spots (in Roman Ingarden's sense). The world it describes, however, does not exist in the material sense beyond the stage proper, for, as explained in the previous chapters of this book, everything that lies behind the fifth wall can only take the form of a mental construct in the mind of the spectator. Thus, the actual words uttered on the stage denote some fictional language that the figures use. At the same time, however, all the linguistic utterances unavoidably refer to what is seen and heard on the stage, which usually is to a great extent incompatible with the implied meanings beyond the fifth wall. Hence language, as used in theatre, reveals the features of theatricality, that is the duality of its orientation. The duality of language is best observed in its referential function: as indicated above, all words refer – via the figures – to the fictional sphere and simultaneously to the phenomenology of the stage (and, indirectly, the world outside), and the latter includes also the words themselves.[5] In our focus on the substance of the signalling matter, we must not forget that words are also part of that substance. They always relate one to another. In this way at least three layers of meaning are constructed: the linguistic (words constitute a dramatic work), the fictional (words denote the language that the fictional figures use and through that may refer to the fictional world), and the phenomenal (words refer to the material substance of the stage). One ought to add referential, topical and intertextual meanings to the list. This leads to a complex network of equivalences and relationships, not to be found outside theatre.

The rules that apply to language on the stage go far beyond grammar,[6] and beyond the rules of literature, for language is subject to the rules of theatre without which it becomes, to use Shakespeare's own words, "a sound, but not in government" (*Midsummer Night's Dream*, 5.1.123). The role of the director in theatre is to bring all the elements of the production under his or her "government", a rule different from that of grammar or literature, to sculpt the language. The language on the stage becomes inseparable from the actors' face and body and voice, their make-up, wigs, costumes, gestures and movements, and all of the latter create a network of relationships, unique for a given production, contained within the time and space given, along with the set, music, lights, space and the like. In many ways this may be seen as an example of transmutation, whereby various elements of the verbal text (not only the stage directions) are "translated" into actors' gestures, costumes, elements of the set, music, dance and so on. They cease to be an acoustic phenomenon only, becoming iconic. For that reason, the definitions and rules of language presented by linguists do not fully apply to its peculiar usage in theatre,[7] for a number of rules for the latter are created *ad hoc* for a given production and are unique. Also, as proven by cognitivist psychology, perceptual understanding is not at all linguistic; also, visual and linguistic knowledge are not the same things.[8] Thus, when we hear the word "ghost" or "apparition", as in *Hamlet*, we do not perceive and memorize it in the same way we would in the world outside theatre, but we blend the word with the manner it is visualized on the stage, and also with the implied meaning or function it might reveal in the fictional realm. Also, we must keep in mind that the spectators view the stage with two oscillating modes of perception: when viewing the inanimate components of the signalling matter, visual perceptions are at work, when, on the other hand, viewing the actors at work, the "visuomotor representations" are generated.[9]

Let us observe that at every moment of the performance every word enters into relationships of varying degree with every other word spoken until that moment and following that moment, with every word spoken by the same actor and other actors, even if they never appear together in the same scene (paradigmatic equivalences are created irrespective of the actors' configurations). This is the result of the literary selection and arrangement of words, their modelling as a piece of drama. It is worth noting that the rule of combination in theatre is of particular aesthetic importance and is conspicuously different from that in literature or language. I have already mentioned in Chapter 1 that the contiguity of particular elements in theatre is not only linear but also spatial, for they appear on the stage simultaneously, creating three-dimensional "pictures" or pictograms. Thus, the rule of combination in theatre governs also the composition of an individual picture, whereas the rule of selection points to the equivalences between this and other pictures, of which the whole performance is constructed. However, in theatre the situation and relationships become more complex, because, as mentioned above, every word becomes inseparably merged with its immediate material context and in this sense is unique to each production. This means that at any moment of the performance, every word enters into linear relationships with all words preceding and following it (as is the case of the printed text of drama), but it also enters into spatial relationships with the signalling matter, human bodies, objects, music, light, etc., simultaneously present on the stage. In this case, it becomes a component of a spatial sign and creates meanings within a single stage "picture". Moreover, this particular feature of a theatre performance enables the appearance of the characteristic tautologic mode of expression: different components belonging to different categories of objects, phenomena, along with the human body, acting, language, gestures and movements, may simultaneously refer to the same signified beyond the fifth wall. The negative aspect of the tautology effect is weakened by the fact that the signifiers belong to different categories; instead, we are dealing with a varied and more powerful message, an emphasis, and the director may experiment with various selections and combinations of the signalling matter. But the evolving stage pictures create meaningful sequences; so do the words, thus becoming temporal signs. So, every word or a whole utterance may be related to a single unchanging stage picture, and also to the evolving sequence of pictures. Therefore it does not suffice to say that there are two axes, the syntagmatic and paradigmatic, along which words appear on the stage. Each axis is dual in nature, and may be related to the phenomenology of the stage and to the fictional realm. This shows that in theatre we are not dealing with natural language,[10] and for this reason we should perhaps distinguish *stage speech* from language. Stage speech has a complex heterogeneous substance, comparable to an amalgam, for it is not only an acoustic articulation of phonemes, but incorporates also, through spatial and temporal contiguity, various bodies, objects, light and phenomena that appear simultaneously with it. *One does not only listen to stage speech, one also watches it.*[11] In theatre we are dealing with "speaking pictures" – to use Ben Jonson's phrase. And since the heterogeneous iconicity of individual words and utterances is unique for a given production, the language employed loses its discrete units, becoming part of the new system which is indiscrete. In other words, the dramatic text provides the material for the verbal utterances of the actors, whereas stage speech, being of composite substance and modelled differently in every production, becomes a component of a heterogeneous sign of the natural language that the figures use, the language we cannot hear and which is only implied. It is also from other signals provided by the actors, and by other components of stage speech, such as props, music or light, that we come to understand what meanings the utterances in that language denote in the fictional realm.[12]

Naturally, the dramatic text in itself provides much information, not only concerning the plot; it also gives us insights into the figures' psyche, explains stage business and describes the props, scenery, music, etc. In the case of verse, additional information is conveyed by the rules of poetics. As a piece of literature a dramatic text also reveals a conspicuous aesthetic function. But, again, many of the pieces of information conveyed by the text, and its aesthetics, appear to the audiences only, and are beyond the perception of the fictional figures. Let me give just one example. When the language used on the stage has the character of a poetic utterance, the mimicry, movement and gestures, all combined to form the text of the figure created by the actor, will be partly subordinated and attuned to the features of the verse. Most often this takes the form of stage action or acting, which are a tautological strengthening of the verse metre, the rhythm or imagery. The poetic text becomes therefore the main stream of the actor's scenic articulation, and the remaining components of the signifier are appropriately selected and modelled in such a way as to create equivalence in this, and no other, arrangement. Thus, not only is the "rhythm" of the gestures or movement of the actor correlated with the verse metre, but also a break in the course of the line (for example at a caesura) becomes significant, particularly if it is part of a dialogue and the second figure does not complete the missing feet in the broken-off line. Rhythmic equivalences inevitably transform themselves into semantic equivalences. For example, in *Hamlet*, one of the many scenes in which this phenomenon can be seen is the one containing the dialogue between Polonius and Ophelia (1.3). The girl admits that the Prince has recently been paying considerable attention to her:

> He hath, my lord, of late made many tenders
> Of his affection to me.

And here she breaks off after half of the line (totalling five iambic feet, that is ten syllables). In a play written in verse this often happens, but there is a way of maintaining the rhythmic flow – the second figure simply "comes into" the given line and adds the missing feet. In this particular case we are left with the space of three syllables, which is easy to fill. This, however, does not happen here and Polonius replies, beginning a new line of verse:

> Affection? Pooh, you speak like a green girl,
> Unsifted in such perilous circumstance.

Shakespeare could of course have written:

> *Ophelia:*   He hath, my lord, of late made many tenders
> Of his affection to me.
> *Polonius:*   Affection?
> Pooh, you speak like a green girl, unsifted [etc.]

But he did not do this. Why not? Because the break in the verse and the missing syllables indicate the surprise or distress which these words evoke in Polonius. He quite simply does not know what to say and so he gathers his thoughts and seeks the appropriate words (the number of missing syllables becomes a measure of time; obviously in the theatre time is relative and needs not be related to the actual time required by the demands of the verse flow). This can be – and usually is – clarified by mimicry and gesture, movement or its lack (a sudden freezing by the figure). But let us notice here that nonverbal signs acquire an unexpected meaning simply because they have been written into the superordinate structure of the verse. We see therefore that equivalences of signs from different systems have been superimposed on the principles of the structure of the verse and in its stream they create a specific meaning. The verse, in turn, can be rhythmically co-ordinated with the music or the lights, constituting further subordinate substances of the "reaction" within the ambit of the signifier or an amalgam. The director may, of course, decide to delete some of the words, thus destroying the verse and annulling the conveyed meaning; he/she may also decide to have Polonius cut Ophelia's line short, and continue without a pause of surprise: again, the meaning of the scene will inevitably be different. It is therefore important to remember that meaning is generated not by the language alone, but by a network of relationships between heterogeneous components reacting constantly and in changing configurations on the stage. As stated above, every word articulated on the stage enters into at least three layers of meaning: the linguistic, the fictional and the phenomenal.

Speaking pictures

As we can see even from these introductory remarks, the language in theatre is a topic in itself which still needs a thorough theoretical scrutiny in its own right, especially in relation to its transformation on the stage: for me it is apparent that the language used on the stage during a particular production is not the language in the drama that Shakespeare (or any other playwright) wrote. Phonetically and syntactically it may be the same, but does not have to be, as in translations; in many cultures the phonetics of the language used by the actors on the stage is different from the every-day style. This is not only because the language of a play in performance is not a natural language, and it is not an isolated linguistic phenomenon. The essential change lies in the fact that all the components of the performance become subordinated to the rules of the medium of the theatre, thus signalling the appearance of a superordinate structure (i.e. a theatrical performance). The same happens with words on the stage: we still recognize them phonetically, we acknowledge their usual meaning, referential and poetic function, but they also enter into various "reactions" with other words and the visible non-verbal elements of the production, creating all sorts of amalgams or clusters that constitute a new "language", basically iconic and heterogeneous, that we, the spectators, have to decipher. Since the signalling matter varies in every production, the stage amalgams, and hence their meaning and the hierarchy of functions, are also different. This undermines the validity of the so-called isomorphic relationship between words and their meanings.

Thus the full meaning of verbal utterances is the relationship between the realms divided by the fifth wall. For this reason also, it is worth repeating, language, although autonomous, is not independent, since it creates compounds, chemical amalgams, with other non-verbal components, from which the director, designer, composer and actor create the performance (their communiqué, or artistic event). These heterogeneous amalgams are theatre signifiers, and they differ fundamentally from the signifiers of natural language or other systems of signification. Moreover, natural languages are composed of discrete elements, whereas theatre as art is indiscrete, for every production is of different chemical composition and is composed of units and reactions that cannot be found and used in any other production interchangeably. If it were otherwise, then there would be no need to put on plays; reading them would be sufficient. This means that seemingly the same words may mean something totally different, depending on the material substance and its modelling of a given production.

Since the stage speech is not only uttered by the actor, but is also signalled by the non-verbal context appearing on the stage (and the stage itself), the theatrical speech-apparatus goes beyond human anatomy. The stage context of the verbal parts most often appears as a kind of spatial, material, although translucent, "retort", which, through its shape and structure, individual features, composition, ability to create variable "frames", "foundations" and spaces for each dialogue, scene and plot sequence, fulfils the function of a specific, developed semiotic "mimicry", further unveiling and foregrounding all the scenic reactions. In the artistic sense, the performance creates a dynamic spatial picture with a constantly changing selection of components and their composition (a completely static performance is hard to imagine), which may be seen as a series of reactions involving different ingredients and yielding different meanings. We can here distinguish between the theatrical "speech apparatus", which is the permanent material frame and basis (with all its machinery and technicality), and the variable substances of the signifiers of fictional worlds that are generated and articulated by that apparatus, which is the stage speech. Even if the scenography does not change, surely the intensity of the light or its colour does, and above all so do the proxemic relations of the actors, their blocking, and the words they utter. Each move changes the composition of the whole. In our mental reception we transform its objective materiality into subjective significance and fictionality (with the latter becoming a mental construct). And the traditional, dictionary meaning of words is contrasted with the actor's interpretation as well as with the material substance and its composition in the amalgams created in a given production. It may therefore be said that the stage speech is a sequence of heterogeneous acoustic and material amalgams, generated in different configurations in every production. These amalgams are both spatial and linear in their appearance. Since there is no common stock of these amalgams, the originality and talent of the director is revealed by his/her ability to combine different components in such a way that their reactions to one another will generate meaning. The latter, of course, always derives from the relationship of the implied meaning, denoted in the fictional sphere, to the modelled and structured phenomenology of the stage. And the artistry of a given production depends on whether it can explain its appearance in the form given and justify the components employed. In rare instances, the scenic text reveals a tendency to denote just itself, being totally, or nearly totally, self-referential, as is the case of some avant-garde, or postmodern productions; this may also be an example of art for art's sake.

Furthermore, it may be observed that the language in theatre does not serve natural communication between people. As pointed out above, the real addressee of everything which is said during the performance, is the spectator, and not the interlocutors engaged in a dialogue on stage. Thus, the dialogic situation in itself is not natural, but predetermined and imposed upon the actors. It is a fulfilment of their professional duty and not a natural act of communication. At best the ensuing dialogue is partly a sign of an interaction between people doing their job. It is the playwright and the director who "speak" through the voices of actors, who represent stage figures, and through everything else that can be seen or heard on the stage. The true axis of communication is therefore between the creators of a given production and the audience, the implied director and the implied audience, if you will. The actors communicate with each other only in the sense of signalling, verbally or not, their cooperation in realizing the artistic goals. For this reason the signalled communication between actors can at best be a sign of interpersonal communication between fictional figures but not an act of natural communication itself. We are not eavesdropping in theatre, we are not watching people and events that are independent of the limitations and rules of the communiqué created on stage: everything we see and hear constitutes the encoded message that someone is sending us, for example, showing us the sufferings and pangs of rejected love, the proud man's contumely and the slings and arrows of outrageous fortune. It is not the actor who suffers, but the actor's fictional referent that we can neither see nor hear (i.e. the "character"). The actor indeed may feel not pain, but great joy and satisfaction in performing the sufferings of the tragic hero on the stage.[13] Also, the levels of awareness, providing the cognitive context necessary for the understanding of any verbal utterance, are distributed freely and purposefully by the playwright, the actor and the director. The context not only differs between the figures and the audience, but also varies from one figure to another, depending on whether or not they witness certain scenes and dialogues; and this may be contrasted with the audience's awareness, based on scenes that the spectators have witnessed and some of the figures have not. This is also a way of creating suspense and so-called dramatic irony. The actor's awareness and emotions are something else altogether, and are not part of the message.

As mentioned above, on the stage every word creates all sorts of equivalences not only within a given dialogue, but also with things said in dialogues occurring in other scenes, with the participation of other figures (and also with non-verbal signs). Obviously the figures are not aware of those equivalences and do not know that they are being "eavesdropped" upon (since the figures do not know anything beyond what the playwright, the actor, and/or director wants them to know), and they do not know that their particular utterances create a dramatic and theatrical structure. The creators of the performance know this very well, as does the playwright, who constructs the dialogues in such a way that every utterance on stage can be meaningful also for the spectators (who may recognize the dramatic structure of the verbal utterances) and can deliver in a relatively short time as much information as possible to the audience concerning the where and when of the action, explaining who is who and what drives the conflict and the plot. What is important here is that this information has to be conveyed in an indirect manner, as if independently of the presence of the spectators, of whom the figures are not aware. This is why dramatic expositions are very difficult to write and are often marred by redundancy, as for example when the information conveyed in a dialogue is obvious to the interlocutors. Since it is obvious, it is redundant, and we, the spectators understand that the information is in fact addressed to us. By this means the speakers' awareness of our presence is signalled and acknowledged, and this in turn violates the rule of accessibility. The playwright shows his/her incompetence and inability to provide sufficient amount of information through theatrical means. Moreover, the verbal redundancy invalidates the actor's efforts to signal the separation of the two worlds. The effect of theatricality is weakened if not destroyed.

Thus, what is created on the stage are indeed "speaking pictures", three-dimensional installations, moving spaces, of which the articulation of language constitutes just one of several inseparable components. Just as in musical compositions a human ear will distinguish particular instruments and voices, so in theatre we become aware that we are in fact dealing with "chords" of heterogeneous components, of which the human language is just one. However, again as in music, its function and meaning will be determined by the signalled relationships, or reactions, with all the remaining components. The selection of the latter, their composition and their sequence will naturally be different in every production. As noted above, there is no common stock of scenic amalgams available to directors: every time, in each new production, they have to select and combine components in a way that nobody has done before. It is obvious, therefore, that the selection and creation of individual components is not the end in itself – what counts is their combination, their spatial arrangement in individual "pictures", and in sequences of these pictures.

Again, what is important here is the established relationship between the two functions: an actor's utterance treated as an articulation of natural language and as a component of stage speech. Each of these means different things, hence another relevant relationship is that between the meanings established in the world outside, in the fictional sphere, and the meaning in the phenomenology of the stage. I have already noted that even the phonetic shape of an utterance on the stage does not have to be congruous with that in the fictional world. This is immediately noticeable in translations. An Armenian production of *Julius Caesar* is a sign of Early Modern English of the original play, which in turn, is a sign of ancient Latin. Similarly, a linguistic utterance on the stage does not have to be a sign of any verbal activity on the other. Loud utterances by the actor-spectators uttered during a play-within-a-play may be a sign of just a whisper in the fictional world. The silence of an actor may be a sign of actual speech of the figure he enacts (as in a scene in *Othello* by Luk Perceval, 2005). In many cases a soliloquy may be a sign of a mental process rather than of actual speech.[14] Eli Rozik has described an example of a production from Israel in which a technique was introduced that is reminiscent of silent film conventions: the actors moved their lips without audibly articulating a sound. The stage speech in this particular production, deprived of the phonetic component, was supported by other components, namely, actors' gestures and mimicry. These became the stage signifiers of the natural language that the figures used. Surprisingly, the stage speech was also understood without much difficulty by the audience.[15] Also, in postmodern productions it is common to detach the voice from the body of an actor, which, consequently, leads to the estrangement effect and de-psychologization of the traditional "character".

The actor alone is not capable of signalling what his/her utterances mean behind the fifth wall (if they mean anything: it may happen that they do not occur there as an acoustic phenomenon). For that purpose responses and signals from other actors are needed. These utterances and signals (body language, facial expressions, etc.) show us what a given utterance means behind the fifth wall. If an actor is saying something aloud and others do not react and do not notice the fact of someone speaking, this means that we are dealing with a convention, with a soliloquy or an aside. It may also mean a concealed intention of the figure, which, for various reasons, is not articulated aloud in front of the others: then the voice of the actor becomes a sign of those hidden intentions, of an inner voice. If an actor is not saying anything in an audible way, but is only playing the harmonica or saxophone, responding to other figures' verbal utterances (as was the case in Krzysztof Babicki's *Anthony and Cleopatra* – *Teatr Wybrzeże*, 1996); or if he/she remains silent, but the others react as if he/she were saying something, then we understand that behind the fifth wall the enacted figure is in fact talking. As we can see, in either case the meaning of the scene will be different, depending on the material vehicle and the modelling of the signifier. Also, it may be noticed that non-verbal signs have the ability to substitute for the signs of a language.

Let us therefore repeat that the words actually articulated on the stage, along with their material co-signifiers, are only signs of the language that the figures use.[16] We cannot hear the latter, because by definition they are fictional, hence occur, by implication, at another time and in another space. The implied utterances of fictional figures have their distinct spaces of reference, through which they create meanings and have the ability to build in the mind and memory of the spectator the fictional realm, the plot, "characters" and so on. But they refer to the fictional, hence immaterial world, and not to the phenomenology of the stage. The language used by the figures does not reveal that quality. It is not the figures who point to a fake stone and call it a "diamond", or sit on a stool that represents the throne, holding a stick that stands for a sceptre. It is the live actors who do that. This is the major difference between the stage speech and the implied language of the figures, which does not reveal the effect of theatricality, with the possible exception of the theatre-within-the-theatre convention. Beyond the fifth wall, words mean the real thing, not an element of the set or a prop. They are not contrasted with anything that the spectators can perceive with their senses. Thus, every unit of stage speech refers to at least two realities simultaneously, to two ontologies, two spaces and times, and it is due to the juxtaposition of the meanings created in both of these spheres that theatre reveals its unique ability to create its particular modes of signification. But to create and blend these two spheres (input spaces) in the mind of the spectator, one needs someone who can describe to us what lies beyond the fifth wall, to clarify the indeterminate spots. This is done by the fictional figure, naturally through the utterances and behaviour of the actor. Once again, we find corroboration for the observations made in previous chapters, where the juxtaposition of at least two models of perceiving reality was presented as a *sine qua non* of theatre as art.

In other words, the language on the stage, preserving its substantial characteristics (with few exceptions, as in the works of Richard Foreman or Robert Wilson, where words are often turned "inside out"), falls into at least two semiotic systems, each governed by its own rules and by the intrinsic hierarchy of a semiotic order.[17] Consequently, one of these systems, verbal in its essential features, preserves the acoustic and systemic characteristics of a natural language, while the other, simultaneously, becomes a scenic amalgam, an iconic sign, a "speaking picture" if you will. We are thus dealing with a paradoxical situation: a simultaneous articulation of two systems of signification, each set within a different time and a different space, and yet both stemming largely from the same source, which is the actor's voice, supported by the signalling matter. We may therefore conclude that actors, paradoxically, speak two languages at the same time: one of these is natural in the sense that it refers to the phenomenology of the stage and to the contemporary world outside theatre, while the other ceases to be an acoustic or purely verbal sign, becoming a component of stage speech. It is therefore justifiable to treat the two mentioned systems as not identical; instead, one is superimposed on the other. The totality of this phenomenon constitutes stage speech, which allows us to talk about the shape of words. We may notice two tendencies here, namely, when the juxtaposition of the two "languages" on both sides of the fifth wall reveals only slight differences, or when the latter are considerable. In the first instance, it is the language of drama that dominates and subordinates all the remaining components, in the second one, it is the substance and its modelling that subordinate the dramatic component. Consequently, the perception of stage speech is different from the ways in which human brains perceive language. This evidently needs further linguistic and psycholinguistic or cognitive scrutiny.

It is important to remember that no natural language is capable of describing the entire complexity of the phenomenal (or mental) world. Even in its most advanced literary usage, language cannot describe all the features of a human face, of an interior decor or a view from a window. And all linguistic attempts to describe the world are linear in nature: they need time to reveal their entirety. Owing to their limitations, all natural languages employ all sorts of stylistic, metaphoric and rhetorical devices in order to convey the desired meanings. However, one of the obvious limitations of natural languages is the fact that they are not capable of a simultaneous presentation of details and complexities of even small fragments of reality. By definition, a human language is not capable of describing anything in the visible or imaginary worlds in a way that is not linear. As a rule, language descriptions concentrate first on the whole and then they move to details. What is not included in the description is (re)constructed by the mental processes of the recipient mind. In theatre, however, the situation is strikingly different: in each temporal unit we call a scenic picture, the spectators are presented with a fragment of reality that contains all its constituent components and details simultaneously (it is at the same time a highly metonymic reality). The fact that the stage picture is often not similar to the world outside theatre, or that it requires time for the spectator to notice and comprehend all the details presented, is not so important. What really counts is the fact that all the information is presented simultaneously. It is a slice, a cube rather, of a material world that can be measured, weighed, described and photographed. What in linguistic description requires selection, here is given in full dimension while the selection in the cognitive process of perception is bestowed upon the spectator, who can freely wander through the multitude of human bodies, objects and other phenomena on the stage. Anyway, in its scenic presentation, the phenomenal world appears simultaneously in its entirety. Moreover, there is nothing beyond it, in the material sense.

In any case, natural languages have a wonderful capacity to describe spaces, actions or objects in their wholeness ("there was an old table standing in the kitchen"), but are much less able to deal with the innumerable details of whatever is being described. A description of a kitchen that included every piece of furniture, every object and utensil would be unimaginably meticulous. One cannot simultaneously say that a human being is laughing, by which its face changes its appearance, that tears drop from the person's eyes while his/her belly shakes and he/she unbuttons the collar of his/her shirt or blouse. This leads us to assert that natural languages treat the whole as more important than the details. This is a systemic feature and not a matter of individual choices. Even the most detailed description fails in providing all the details; these, however, are presented simultaneously on the stage. Consequently, it may be said that stage speech conveys more information than the language alone. A word uttered in a visible and audible context, with which it becomes blended, means much more than outside that context. "A table" articulated in a natural language, if it does not refer to a concrete object, or if the denoted object exists in another space of another time, can have a multitude of different shapes, which may, of course, be clarified in further description; but basically it means a class of objects of specific function. A table on the stage has a given shape and ascribed function. When someone says, "leave this on the table", we know perfectly well which table is meant and what are its individual features. What we do not know are the features of the table denoted by the same word in the fictional sphere. We can notice at once that an articulation of a human language on the stage has a dual referential function: the word's direct meaning is the concrete table, whose features need not a further description, being part of the set (the stage table might be metonymic or even immaterial). But the combination of the one with the other, i.e. the word "table" with whatever is its stage denotatum, constitutes the signifier of the table which is not concrete at all, for being fictional it is by nature immaterial, hence invisible. Thus, we are dealing here with a different system of signification than is the case with a natural language. On the one hand, the function of the stage signifier becomes similar to what the word denotes on the stage, but in a substantial sense it is not the word alone that denotes a particular element in the fictional sphere; instead, it is the word blended with light, music and the voice and body of the actor who articulates it, and blended also with its material denotatum on the stage. All these components of a blend or amalgam constitute the material vehicle of the signifier, being an example of stage speech.

Moreover, the fact of naming an object as a table is significant, for in theatre we may also be dealing with the phenomenon of different substances representing the same objects or phenomena. There may be a real table on the stage, or its metonymy; the table may be painted on canvas or it may be represented by a stool or a human body on its fours. Depending on what represents the table on the stage, the meaning (and aesthetics!) of the word "table" changes. Moreover, the same object, that is the table, may change its functional meaning within the same production. Similarly, an actor may say "a bed" pointing to a table – we then understand that the object means a bed in the fictional realm. The iconic incongruity is part of the meaning conveyed, and may become of particular significance as the performance unfolds. The same object, set upright, with its top set towards the audience, might mean an open tomb, as was the case in Nekrošius's *Hamlet*. A wooden wardrobe in somebody's dressing room may become a mountain that several actors climb, as in the play *Na przełęczy* directed by Andrzej Dziuk (Teatr im. Witkacego, Zakopane, 2002). In the same play, chairs around a table become stones in a stream: the actors jump from one to another, as if to avoid falling into water. In a material context, where there is a particular incongruity between a word and its denoted object on the stage, words (and objects) begin to acquire different meanings from their everyday or usual ones. Through the object, which comes to mean different things, all sorts of equivalences may be created, leading to further complexities of the relationships between words, light, bodies and things, resulting in the semantic and aesthetic richness. Many other examples of this phenomenon are given and discussed in this book.

Thus the articulation of stage speech goes far beyond the verbal expression and reveals its iconic and constantly changing nature. This is why it is reasonable to distinguish between two major categories of material components of stage speech: those which originate in the actor, and those which are spatially external to him/her. The quantitative relationship between the two categories may be helpful in establishing stylistic traits of a given production or a period in theatre history. For instance, in the postdramatic stream, the second category is likely to dominate; in mime, on the other hand, the first one will. Spatial contiguity helps the spectator to determine which of the objects and phenomena appearing on the stage are to be treated as components of stage speech at a given moment. For one has to keep in mind that stage speech continually changes its substance, proportion and significance of participating components.

We must be aware that the substantial form of stage speech in a given scene is independent of its iconic similarity to what it denotes in the fictional world. In other words, we can have a nude actor, sitting on a ladder with a stick in his/her hand and with a recording tape wound round his/her head. All this reacts with the words the actor utters, say, in French, and creates a sequence of stage speech, which does not at all mean that the fictional figure winds the tape round its head, or that it is naked, or sitting on a ladder or speaking French. It may, for instance, mean a state of the figure's emotions or mind, and not its concrete behaviour, habit or speech. Therefore, one ought to agree with Eli Rozik who observed that the substitution of the verbal component within the context of a non-verbal one, will not significantly change the generated meaning, whereas the reverse situation, that is a substitution of the non-verbal context will significantly alter the meaning of words.[18] Truly, the words of Benvolio in *Twelfth Night* will have a different meaning if he is wearing yellow stockings, than if he is not, or if the stockings are not yellow; they will mean differently if he is holding the forged letter in his hand. And yet they will mean differently, depending on the way, in which the actor articulates his words, for instance if the actor signals that his fictional ego is aware of the whole situation. Good actors have the capacity of articulating a single word in a great variety of ways. But let me yet again draw attention to the fact that in theatre it is not only what is present on the stage that constitutes the meaning, but also how all that is perceived by the fictional figures. Benvolio may appear without any stockings at all, and yet, within the fictional realm, all the figures will notice him wearing them. "Deck our kings with your thoughts", advised William Shakespeare.

Language thickened in a retort

The above observations lead to the conclusion that also in the case of language the effect of theatricality and the rule of accessibility are observed. Verbal utterances on one side of the fifth wall do not have to have the same "form" as their equivalents on the other; moreover, they do not have to occur at all, and for that reason will have referents on both or just one side only. Their denotata might be different on both sides of the fifth wall, sometimes strikingly different, drawing attention to the words and objects and the newly established, uncommon relationships between them. All of this means that the stage speech is always modelled by its creators for a particular production, while certain recurrences in different productions may be the trace of the stylistic characteristics of a given director or stage designer, or of a trend or vogue. Equally, they may sometimes serve as quotations, or may even be a sign of plagiarism. Stage speech is always unique for a given production, even in the smallest detail. And this is achieved in ways that are contrary to the rules of natural languages, which in their basic shape and grammar retain regularity and reiteration. This is why we can say what is correct and grammatical and what is not, what breaks the rules, what is vulgar, obscene or poetic.[19] All of this implies that the speech-act theory cannot find its full realization in theatre (as indeed no theory that is based on a natural language can). The actor is an inseparable component of stage speech, whereas the fictional figure is not: it uses natural language, and in the vast majority of cases makes its verbal utterances outside theatre. The actor, with his/her behaviour, gestures, mimicry, etc. clarifies not only the words that he/she articulates, but also clarifies the ways in which the fictional figures use their language, their modality, along with signs of emotional states, relations with and attitudes to other figures, etc.

Thus, it may be said that stage speech is something much broader than the natural language. Stage speech is constituted by a network of relationships between components of various origin and substances, mostly not acoustic and not linguistic at all. And for this reason stage speech cannot be analysed in the way natural languages are. Beyond its material structure, unique for each production, stage speech ceases to exist. This explains also why stage speech cannot be translated into any known human language. It cannot be transcribed. Thus, there is no natural language present on the stage alone, and there is no factual verbal activity that can be described and analysed in all its complexity with the aid of the tools and methods of linguistics. This is why I shall stress once again that stage speech does not refer directly to the material and phenomenal world beyond the fifth wall, but creates meanings in three phases: (1) it denotes the implied verbal utterances of fictional figures, and it is these utterances that are no longer stage speech, but a natural language articulated by fictional figures only, that (2) refer to the material and other phenomenal components within the fictional world; and (3) the ultimate meaning (and aesthetics) is generated by the juxtaposition of the (1) and (2) and the phenomenology of the stage. Consequently, one semantic field (or conceptual blend) is contrasted with the other, and this contrast is the basic method of signification in theatre. This is why the verbal is continually contrasted with the iconic, which forces the spectators not only to listen to stage speech, but also to watch it.

Although I shall discuss the issue of acting in greater detail in the following chapter, it is worth noting at this point that everything that occurs on the stage, including acting, may be referred to two realms simultaneously, the world of the stage and that of fiction. The factual utterances of the actors are not and cannot be true speech acts, because (a) they occur in a situation of total pre-programming and (b) they do not refer to the speakers (that is the actors), because, simply put, they do not concern them. And "theatrical speech acts" are not just "iconic replicas of real ones", as Eli Rozik tries to convince us.[20] In theatre, there are no "real ones" or their exact replicas. As signs of the natural language that the figures use, they may, of course, also be signs of various speech-acts in the denoted, hence fictional world, but, again, they are preordained and have a future that is determined (from the point of reference of the spectators). Thus, implied verbal utterances of a natural language occur only within the fictional realm, and may therefore be treated as speech acts only within that realm. Owing to this ability they can generate additional relationships and meanings within the whole work. But, let me stress again, what really counts in theatre is the relationship between the verbal utterance in the fictional world and the actual stage speech. As hinted above, their substantial appearance and functions may differ considerably. Thus it is entirely due to the signalled behaviour and utterances of the figures behind the fifth wall whether or not a given utterance of an actor finds its verbal equivalence there, and if it does, then how it will be perceived and interpreted by the fictional figures. And vice versa: non-verbal signals emanating from the stage (silence, too) may be interpreted and perceived by the fictional figures as the use of natural language. We can see at once that what J. L. Austin called the "dimension of locution" (by which he meant the referential function of the language) is different on both sides of the fifth wall, and hence the other two dimensions, illocutionary and perlocutionary, will also be different. This is determined primarily by the director of a given production, and may lead to the rise of a third "speech act", which is addressed directly to the spectator. Stage speech is the director's speech act, but, as explained, it is not entirely linguistic. This also means that the interplay of two or more speech acts, different on each side of the fifth wall, may contribute to the creative aspect of the performance.

J. L. Austin himself was very sceptical about the language on the stage, and in fact he excluded it from any serious consideration in his theory. The appropriate passage in his book is the following:

> A performance utterance will, for example, be in a peculiar way hollow or void if said by an actor on the stage, or if introduced in a poem, or spoken in a soliloquy. This applies in a similar manner to any and every utterance – a sea-change in special circumstances. Language in such circumstances is in special ways – intelligibly – used not seriously, but in many ways parasitic upon its normal use – ways which fall under the doctrine of the etiolations of language. All this we are excluding from consideration. Our performative utterances, felicitous or not, are to be understood as issued in ordinary circumstances.[21]

The quoted passage has caused serious problems to all those who have attempted to apply the speech-act theory to theatre studies. In point of fact, we can find scores of different interpretations of the passage, often ludicrous, by which their authors strove to amend the clear meaning of Austin's own words. The paradox results from the fact that Austin clearly excludes stage language from consideration, while his followers, on the contrary, try to include it. It seems, however, that the great scholar was right: the performance language is not an ordinary or natural use of a language, and does not take place in normal circumstances.[22] One could of course discuss the applicability of such concepts as the "parasitic" use of language. In any case, the quoted fragment has been the cause of numerous controversies and polemics, including philosophical ones, that have engaged the minds of such thinkers as J. Searle and J. Derrida. Thus, one ought to be cautious when applying linguistic theories to theatre.[23] Even a valid conclusion in the case of one production may prove fallacious in another. What may be confirmed by the language the actor uses, will be invalidated by the implied words and actions of the figure. Speech acts will appear differently on both sides of the fifth wall, or will appear on one side only or not at all. Thus, once again, we are dealing with a juxtaposition between an utterance function behind the fifth wall and its phenomenal counterpart. This may, of course, provide an opportunity for creative games with the spectator, which may be partly explained in terms of the speech-act theory. But let me stress yet again, stage speech is not a natural language. It is semiotically "thickened", as Yurij Lotman put it.

Glossolalia

Jiří Veltruský has drawn our attention to the fact that as a sign of a sign, theatrical speech can blur or even eliminate the referential function of the fictional figure's speech.[24] He gives examples of some English mummers' plays, in which unintelligible speeches are produced, signifying the fact that a figure speaks but not what it says; the same device is used in some theatrical folk customs in Bohemia, and in the "artificial language" used in the midst of a thirteenth-century Nativity play from Rouen. Also, we should take into account reverse situations, when non-referential speech of the actors is treated as meaningful by the figures. So, it is possible in theatre to signal the fact of speaking only, with the omission of the referential function in the fictional realm.[25] This in fact has become a common feature of avant-garde and postmodern theatre. There are other devices of this kind, such as, for instance, *heteroglossia*, and, usually, their function is to draw the spectators' attention to the fact of someone speaking (although in the case of heteroglossia the functions of "mixed" languages are much more complex).[26] I have discussed stage silence, being a component of stage speech, elsewhere,[27] so now, to conclude this chapter, I shall turn to a somewhat neglected stage speech phenomenon, namely, to glossolalia.[28] This may be defined as a phonetic imitation of a natural language, which, being "sound without government" signifies nothing in the direct referential way. It is seldom employed in drama, but occasionally appears, as, for instance, in William Shakespeare's *All's Well That Ends Well*, in scenes 1 and 3 of act 4. The cowardly and false Parolles is ambushed by a group of soldiers who pretend to be foreigners. They speak a fake language, invented *ad hoc*, and employ an interpreter who makes the communication with the captured man possible. This is an example of their speech:

Second Lord:	*Throca monovous, cargo, cargo, cargo.*
All:	*Cargo, cargo, cargo, villianda par corbo, corbo.*
Parolles:	O ransom, ransom! Do not hide my eyes.
	[They blindfold him]
Interpreter:	*Boskos thromuldo boskos.*[29]

Before the ambush, the soldiers agree that they will speak in some horrifying language, and that it does not matter that they do not understand one another, because what is important is that Parolles is convinced that they do not understand his language. This is why an interpreter is needed. In this way the spectators are informed about the intrigue. Thus, what we are dealing here with is glossolalia, which means nothing on either side of the fifth wall. In film and in theatre a reverse situation is possible, when we have, say, representatives of an alien civilization (or animals, babies, etc.), who use a language that is not understood by their interlocutors (and spectators), and yet they understand that the language is a natural one for those who use it. In this situation, we know that the language used (awkward though it seems) is natural in the world of fiction, at least to some of its inhabitants. So, the situation is semiotically and semantically different from the one in *All's Well*, where the verbal element of stage speech is not a sign of a natural language of the fictional figures. This does not mean, however, that glossolalia does not make sense. In the example given, its function is to show to Parolles that the soldiers are alien enemies and that they do not understand what he is saying. Problems in communicating with the aliens heighten his sense of peril and force him to betray all the military secrets. At the same time, glossolalia plays an important role in creating comedy (except for Parolles, who is frightened to death). Further, glossolalia in Shakespeare's play is an example of a quasi-language that has at least one unique feature: we may assume that phonetically it sounds the same to the spectators and the fictional figures.[30] It reveals this feature only because no one is in fact communicating in this language. For this reason it becomes a sign of itself, and as such is one of the few components that can exist on both sides of the fifth wall, with the two semiotic orders overlapping, hence generating similar, though never the same, meanings.

Somewhat different, although related, is the use of language sometimes met in recent, postmodern or postdramatic productions, where the phonetic quality of the words is preserved, but their grammatical arrangement is not. Consequently, verbal utterances, deprived of organizational rules, lose the semantic coherence. We are left with sounds, recognizably belonging to a given language, but they are not, to use Shakespeare's expression, in government. The words convey meanings, but their clusters do not. Again, the important thing is if these utterances make any sense in the fictional realm, or, perhaps, they are not perceived there at all. It is also important whether the actors signal the figures' intention to communicate with one another, and the ability of the figures to comprehend what remains incoherent to the spectators.

7
Sculpting the Body, or Embodied Time

Cause or effect?

As indicated above, acting is the most important component of theatre as art, and of stage speech in particular. This is why it is important to discuss it within the context of the latter. However, while being recognized as an inseparable component of theatre art, acting presents an immensely complex theoretical problem. The vast majority of publications on the subject of acting are not theoretical at all, and usually deal with the psychology of the actors, their professional cooperation with directors, methods of training, not to mention all sorts of memoirs, diaries and interviews.[1] In what follows, none of these topics will be dealt with, and the focus will fall on the ways in which acting becomes a TSGC, and a meaning-generating and distinctive factor in theatre.[2] The widespread impact of the Goffman–Schechner theoretical fallacy, by which acting becomes an all-inclusive term, used to describe all sorts of social behaviour and interaction, has been sufficiently dealt with by Eli Rozik, so there is no need to enter into the discussion.[3] The degree of perplexity is deepened by a wave of new streams in theatre and criticism, postmodern or postdramatic, by which the traditional understanding of the nature of acting is undermined or even invalidated.[4] The appearance of performance art, and its impact on the theatre, has brought a large measure of confusion as to the boundaries between different artistic mediums. Instead of alleged pretence, which has become a compromised notion, we are now led to believe that the new theatre relies on the real presence of a human being, which does not create fiction and presents itself as itself only. Thus, we are expected to believe that the body on the stage means the body of a live human being only. It does not carry a sign function, and consequently means only itself, as in body-art.[5] The same applies to virtually all other components of the signalling matter. On the whole, it seems that a particular stream of contemporary art is entering into theatre, and in an attempt to colonize it, tries to invalidate the medium's most basic assumptions. Again, the problem is too broad to be dealt with in this chapter.[6] Nevertheless, even these exclusions do not make a theoretical approach to acting much easier. In what follows, an attempt will be made to capture the essential features of acting, seen not as a social interaction, but as a process of creating theatrical, and hence artistic, texts, which reveal the ability to simulate the spectator's experience. Since the essential distinctive feature of theatre is the creation of the fictional present time, which appears simultaneously with the real present, it follows that also acting may be discussed in terms of temporal issues, as this chapter attempts to show.

Admittedly, the actor on the stage is a complex sign, since the actor, among other things, inscribes on his/her face and body a very complicated continuum of dynamic visual and verbal signals, all of which constitute mainly the sign of a figure.[7] Since the actor is a live human being, it appears to the spectator that his/her behaviour and utterances originate inside the actor and are determined by his/her reactions to the surrounding visible and real world outside. This is why the mechanicals in the *Midsummer Night's Dream*, who out of fear of the consequences want to avoid responsibility for the figures' actions, find it necessary to explain to their audience that they are not really whom they impersonate. However, the complication surfaces once the spectator notices the awkward fact that the actor wants to be perceived as someone else, living at a different time and in a different location. This is an obvious contradiction between the biological self, which we see and hear on the stage, and the assumed "I" of the actor, i.e. his/her deixis. Metaphorically, it may be said that the actor puts a new deictic costume on. Depending on the fabric and style, the actor's private deixis may show through the costume, or it may be almost totally concealed. This is what Rozik calls the "deflection of reference", being the distinctive feature of stage acting and resulting in double indexality. Also, Rozik draws attention to the two fundamental sets of distinctions which the principle of acting entails. "The first is between the actor as producer of text, who is part of the real world, and the actor as an inscribed text or description"; the latter is what we have labelled as stage speech, of which the actor's body is just one of the components. "The second is between the actor as the description of a character, and the character as the fictional referent of the description or as a denizen of a fictional world." [8] We should add a third distinction, namely, between the fictional referent and the live human being on the stage, which results in the final blend in the spectator's mind. Please note that through simulation the spectator does not access the actor's emotions and feelings, but the fictional figure's. This means that conceptual blending is essential to the experience of the performance, because it enables the spectator to perceive, at least partly, the actor as the figure, and not only as a person at work. This unique experience makes the figure and its realm more real, so to say.

Let me repeat: fiction in theatre by definition does not exist phenomenologically. There is no King Lear on the stage; there is an actor playing King Lear. But the spectator does not experience the emotions and feelings of the actor; through the process of simulation, he/she has open access to the emotions, pain and sufferings of King Lear. Experiencing fiction becomes a reality. This creates a paradox that lies at the basis of theatre experience. The actor is at least a dual sign, or, rather one signifier with two references. If we use Charles Peirce's taxonomy: as a live human being the actor is an icon representing the fictional King Lear (the general similarity is the key factor here, although that depends also on the ways in which the similarity is acknowledged by the fictional figures), and the stage speech that he/she co-creates as signs of human communication is also iconic; but additionally the actor is an index (here the rules of spatial contiguity and cause and effect apply), by which he/she assumes, through the language, intonation, facial expression and bodily gesture, that the events occurring in the fictional realm have a direct impact on him/her.[9] In other words, the actor signals that events and utterances taking place in a different time and space (and ontology), cross the invisible spatio-temporal border, the fifth wall, and affect him/her directly. Also, he/she signals that his/her own words and actions have an impact on what occurs, by implication, in the other realm. This is corroborated by the utterances and behaviour of other actors. So, the actor appears, paradoxically, as the cause and, simultaneously, as the effect of what we see and hear and what is implied beyond the fifth wall. This is the consequence of double indexality. It is double because the actor as a human being at work is the immediate cause of his/her utterances and behaviour on the stage, and at the same time we are led to believe that these also are the result of and have an effect on someone or something in the fictional world; as we understand, he/she is also the effect of the preordained scenario or script and of the events occurring beyond the fifth wall. I shall return to this paradox shortly.

Thus, what we perceive on the stage is an ontological hybrid, a body of a live human being who, by "deflection of reference", "performs" or describes a fictional being, which also absorbs meaning from other signals, and the spectator is expected to relate one to the other.[10] The full (and individual) meaning derives from that relationship, resulting in the final blend, a resultant structure, which is an aesthetic and intellectual experience. However, the seeming opposition of actor and figure evokes a number of interpretative difficulties. As indicated above, the actor is a co-creator and material substance of the sign, and for that reason equips his/her body and face, gestures, movement and verbal utterances with a sign function, which has two directions (of varying strength, one may say). The one aims at creating an image of a fictional figure in the spectator's mind; the other draws the latter's attention to the verbal and kinesic activity of the actor him/herself that are employed to fulfil the task. This implies, let me repeat, that the ultimate or total "meaning" of acting is not the figure as such – this is just a conceptual input space in the spectator's mind – but its relationship to the physicality of the actor, his/her modelling, characterization, costume, acting skills, etc., and, of course, to the other elements of a given production.

The goal of acting therefore is not only to create an illusion or imaginary construct of another human being (sometimes an abstract allegory or an object), but also to present the ways and techniques of the act of creation. Thus, the true "meaning" and significance of acting (as an artistic endeavour) is not only referential but also aesthetic and experiential, a combination of the two dominating functions. And one more thing, it is important to note that there is also an additional function of acting, which is to assign temporal and spatial dimension and structures to the signalling matter outside the actor's body. It is the actor who splits the stage time into two streams, which, through convention, share only the present moment. Thus, the actor may be seen as the central Time- and Space-Generating Component (TSGC) of every theatre production. Moreover, he/she generates and clarifies the fictional space, as discussed above, for instance, in the context of the thrust stage. No object or animal has the inner ability to achieve that without the support of a human agent. This is why without a live actor no theatre is possible. It is the actor who activates objects and through ostension, realizing the artistic premises of the director, selects the components of stage amalgams, the heterogeneous signifiers. So, it is obvious that the functions of acting go beyond the mere creation of a fictional figure. The actor creates the fictional present time and is therefore the key creator of the theatricality effect and of the stage speech, with the latter being of ultimate importance.

In any case, it is not enough for the actor to signal that he/she is someone else; the actor has to clarify that the presented fragments from what seemingly is his/her life are taking place here and now, in the space contiguous with that of the audience, without, however, recognizing the theatrical nature of everything that is being shown and said. Thus, through spatial contiguity objects, animals and other components of the signalling matter absorb from the actor the time and space that he/she inscribes on his body and through his/her utterances. And it is only then that they may "act", as Rozik wants us to believe. Without the indexical relationship with the actor, they remain incapable of creating fictional time structures in their continuum, and cannot act out of themselves. At this point it may also be observed that a number of pieces of information conveyed by acting are inaccessible to the figures (who, for instance, are usually not aware that anyone is acting). As is the case with other theatre signs, also signals emanated through acting may be divided into those which are transmitted through the consciousness of the figures, and those which are not. The former constitute the indirect channel of communication, which is essential to theatre as art: it enables the appearance of the convention, by which theatre evolves as if out of itself, pretending not to be anyone's creation. It wants to be perceived not as a narrative about events past, but an evolving present, congruous with the present time of the audience. But the direct signals of acting point to the technical side, the skill with which an actor fulfils the artistic tasks, and for this reason they play the key role in the aesthetics of a given production.

Peter Brook has once said that "all that is needed for an act of theatre is a bare stage, an actor, and a watcher". Brook, of course, is not a theorist, but a practitioner, and a practitioner of genius, and when he says something about theatre it cannot be dismissed just because he does not use critical jargon, or does not define the words he uses with great precision. He does not create a treatise that is logically and internally fully justifiable and self-explanatory, as theorems tend to be. The key word here, I think, is the "actor" whom Brook brings to his bare stage as the most important component of staged art. The word itself, as used by Brook, contains in fact the gist of stage acting: the appearance of a human being who behaves and talks in a strange way, rests on signalling to the "watcher" (i.e. spectator) the reasons for the strange behaviour and odd utterances that we witness within the space of the stage. The space of the stage can be "bare", because it is the actor who clarifies its meanings. What is important, however, is the very word "stage", because it implies that the actor has to find him/herself within a space that indeed is a "stage", so the demarcation has to be drawn between theatre and non-theatre. The stage is a very special space, it is a chemical retort, where sundry acts of theatricality are produced. Furthermore, the meaning of the word actor, as used by Brook, does not denote someone's occupation, but an act of artistic creation – and in this sense someone becomes an actor only when behaving in a certain way and only when that behaviour is accepted as an act of communication governed by specific rules that are different from the ways humans communicate in their everyday lives. And Brook knows precisely when that happens, when the behaviour and utterances of a person are of the kind that cross the boundary between the real and the fictional, "ordinary" and "artistic", when they reveal the feature of theatricality, when they put their deictic costume on. This particular behaviour makes someone an actor, and it is precisely that which constitutes an actor as understood by Brook.

So, in a private meeting X may be an actor only by occupation, but at that moment, which is part of his/her own everyday life, he/she is not an actor in the artistic sense, even though some aspects of his/her behaviour and qualities of voice we may classify as "theatrical". When the same person appears on the theatre stage, however, and, dressed perhaps in an odd way, with or without heavy make-up, perhaps wearing a wig, or in his/her seemingly "ordinary" appearance, begins to behave in a peculiar manner and to utter strange words that seem to belong to a different context, different time and a different person, we may classify the person as acting (in the artistic sense) and the person as an actor creating a stage figure. This is Brook's "actor". Of course, the qualities mentioned so far do not touch the most important feature that constitutes an actor in the theatrical and artistic sense: he or she has to signal that s/he does not perceive the presence of the audience; this is not done out of rudeness, but to signal that the human being we observe ceases to be him-/herself and begins "playing" someone else, a fictional figure, set in a different time and in a different space. The latter explains why the actor does not see the audience. We have to keep in mind, however, that the figure is much more than an actor capable of creating on his/her own: as indicated on several occasions in this book, the fictional figure is a synthesis of the various relationships between fictional and material and verbal substances of the performance. What this means is that the figure is constructed not only by what the particular actor who creates it does and says, but also by other factors, such as other actors' utterances, behaviour, costume, make-up, light, music, and the like. Not taking this factor into account makes any discussion of acting and theatre rather superficial. After the performance, the same person ceases to be an actor, and returns to his/her ordinary ontological status of a human being whose profession or hobby is acting. The actor ceases to function as a sort of artistic text, which we can "read" as a complex cipher composed of a human body, its gestures, mimicry and voice, but also blended with objects like the costume, wig, makeup, props, etc. This is what Brook means when he says that all he needs is a bare stage and an actor and someone to watch. Moreover, once the acting has started, the created fictional time of the actor's body spreads on everything that he/she looks at, stand on, touches or refers to verbally. The time of the body is absorbed by the bare stage, which ceases to be just a stage – it becomes a space set in a specific time, clarified by the actor's utterances and behaviour. If the actor signals that his time are the roaring twenties, so the stage moves back in time to that period. The same applies to the costume and props, if any are used: they also absorb the time, and thus become theatrical. Time is the meaning. The same rule applies to space, which is also clarified by the actors.

Secondly, the very fact that someone is a watcher, and not a chance passer-by or witness, explains why he or she is prepared to alter his/her ordinary way of perceiving reality, and start reading the performance as if it were a three-dimensional text. An actor makes sense only when perceived in a unique way, when "read" and not only observed in the usual manner we observe human beings. As stated elsewhere in this book, theatre needs at least two real retorts for its reactions to take place: the material stage and the mind of the spectator. The mind of the figure cannot be taken into account because it is not real; it can only be described, or implied. Thus, when Brook mentions a "watcher" he means a conscientious spectator who is "reading" the actor on the bare stage. And in order to start the cognitive process of watching, perceiving and experiencing a theatre performance, the spectator has to notice certain characteristics of theatre that make the phenomenon distinct from non-theatre, has to catch some systemic markers known to him/her, which will cause an appropriate perception strategy to switch on. Also, the spectator brings to the theatre his/her sensitivity, intelligence, knowledge, presuppositions and memory – all of which contribute to the ways the performance is perceived. For instance, the memory of the previous roles of a given actor may influence the spectator's evaluation. The knowledge of cultural codes is equally important. One of the important markers is the fact that the person, say, a man crossing the stage (this being Brook's example), behaves seemingly in a natural way and yet does not notice the presence of the audience and of other technical, spatial, and material details of the stage. And thirdly, the bare stage indeed is all that an actor needs, for theatre is a conventional art, and usually does not pretend to be a mirror image of reality known to us from experience. With the exception of the body and mind of the actor, and a bare stage, it does not need any material substance for the construction of fictional realms. And one more thing: since the actor signals the appearance of dual temporal structures, and the theatricality effect begins to operate, it is crucial that the spectator recognizes this as an intentional invitation to take part in an act of uncommon communication and unique experience. The cognitive process is ready to begin. This requires a drastic change in cognitive receptors, by which the spectator begins to read the actor and the signalling matter on the stage as a performance text. Consequently, he/she does not see a human being only, but an actor who is at work, creating a figure that is not him/her. As the result of stage reactions, not only human bodies, but also trivial objects, common words and mundane behaviour thicken in their semiotic appearance, and reveal their capacity to co-create texts of unbelievable semantic and aesthetic richness. The transformation of the ordinary into the theatrical converts it into a carrier of information of unusual potential. And it draws the spectator's attention to the very process of transformation, the reactions, formula and participating components. A theatrical space is demarcated and everything enters into chemical reactions.

As indicated, fictional realms are the result of mental and experiential processes activated by the stimuli emanating from the stage and are partly determined by the individual features of the spectator. The most important and essential source of these stimuli is acting. An actor does not need a stage set, props, or costume conspicuously different from everyday clothes (in theatre nudity is also a costume); he/she does not need a raised stage, or electric lights, or music. He/she alone is absolutely sufficient to create fictional time and space. This is the most rudimentary distinctive feature of acting. All of this creates a paradox, typical for theatre, which the recipient must solve and accept in order to be able to understand the encoded message of the performance. This is why in my definition of theatre, I speak of "mutual agreement" of the actor and the spectator.[11] A politician at an election meeting may "pretend", may "act", may inscribe a fictional being on his/her body and in this sense impersonate someone else (an honest person, a devout defender of the oppressed, a sophisticated economist and thinker, etc.), and we may label that as creating fiction, or even theatrical in everyday parlance, but he/she will not be creating fictional present time, will not pretend that is not noticing the people gathered. Similarly, in rock concerts, the performing artists, even if they impersonate some fictional beings, never pretend to be somewhere else and at another time. This is why "acting" in the popular meaning of pretending, is not sufficient to describe acting as the crucial part of the theatre medium. I will go further still and say that *all human behaviour that results in the creation of fictional present time is acting* (although not necessarily an artistic creation). A human being's visible intention and attempt to create fictional time and yet his/her notice-able inability to achieve that is what we call bad acting (here, of course, the criteria are culturally determined): what we see on the stage is not an actor, but a human being whose nonsensical behaviour does not communicate anything in a theatrical sense but the inability of the "poor player" to fulfil the assumed tasks within the chosen medium. So, the actor is "convincing" (this is still a common criterion) not because of the psychology or verisimili-tude involved (there may be none at all since an actor may be playing an object, a tree, a plant or an animal, or an allegorical figure), but predomi-nantly owing to his/her ability to create fictional time structures, which will enable theatrical communication and experience. Through this the actor's behaviour finds justification. All of this needs some further explanation, especially in view of various seemingly theoretical discussions and publica-tions on acting, which, as I said, usually deal with the actor's psychology, methods of actor training or his/her interpretation of the assigned role.[12]

Embodied time

Since the live actor, objectively set in his/her biological present time, but wearing a deictic costume of someone else's, through various ostensive signals establishes spatial contiguity with the material substance of the stage (the signalling matter), it is inevitable that an indexical relationship is also established, and that relies largely on its spatial and temporal aspect (i.e., spatial contiguity is established along with a chain of cause and effect, which seem to link or even blend the actor with the fictional figure and the realm it inhabits). We see the two spaces and the two times blend as if in chemical reactions. This is the "deflection of reference" and explains why everything the actor touches, everything his/her gaze rests upon, and his/her words describe, also becomes a sign of analogous situations, objects and utterances in the fictional realm, which, as explained above, in a material and semantic sense do not have to be similar to their stage signifiers. Also, everything we see and hear on the stage becomes indexically linked to the events and actions in the fictional world, beyond the fifth wall; they merge in the cause and effect chain of events. Tears on the actor's face, which are objectively set in the present time, become the effect of their cause, which is the pain of the fictional figure enacted, set – by implication - in some temporal past, say, in ancient Troy. The temporal distance is annulled by the signalled spatial and causal contiguity. At the same time we know that objectively tears are the result of the actor's training or some technical trick. Of course, in theatre the actor can cry without tears; the tears might, however, emerge in the fictional sphere, on the figure's face, as signalled, for instance, by a handkerchief, or verbal description. Thus, breaking through the fifth wall is the essential factor, which enables the medium's modes of signification to operate. However, we must not forget that also the "analogous situations", mentioned above, do not have to be similar to what we see and hear on the stage. They are analogous predominately in their temporal and spatial contiguity or simultaneity. Also, through simulation the spectator may experience, at least partly, the pain, trauma and emotions of the figure that are the implied cause of the actor's tears. The spectator does not and cannot experience the actor's emotions (with the possible exception of bad acting, which, may, indeed, be a painful experience).

Since spatial contiguity and self-referentiality are a feature of an index, what the actor performs on the stage is received as a direct image of the activities of the actor, or the figure,[13] because of which cause and effect coalesce in an ontological and chronological unity and chain that make us willing to submit to the temporary illusion that the actor is the figure, and that the expression of love or anger on his face is the result of real feeling. Whereas in fact its cause lies on the other side of the fifth wall and in the artistic goals set before the actor. We then read the actor through two time structures that merge into one – theatrical – time. But this is not a continuous process, because we at once discern the "skill" of the actor, that is the techniques, actions and utterances that create the figure, a person at work, because of which the actor again appears to us as the cause of what we are watching. Now we discern one time structure – the real one. However, when the actor and his actions appear as the effect of events that we do not see, but which can be brought into existence in our spectators' minds, then the conventional temporal and spatial structure of the theatre's fictional world is signalled once again. And this, as I have said, reveals an elastic time that can be modelled almost at will. As a result, there is a logical and physical contradiction of two present times, but when the acting is good, this contradiction is temporarily suspended and blurred, and we can again submit to the illusion or belief that what we see is the effect of the actions, emotions and "mentality" of the figure, which in this way becomes a being that has apparent (mental) reality and in the European tradition constitutes the basis of the theatrical experience. It seems therefore that the conceptual blending theory is heading towards a scientific explanation of the described phenomena.

This explains why the relationship of the signified to the signifier is so important in theatre, because it draws the spectator's attention to the selection, combination and modelling of the signalling matter used. It constitutes the "missing link" in many theoretical approaches to theatre. This also explains the misunderstandings originating in the interpretations that take into account one side of the fifth wall only. For instance, Jiří Veltruský, when analysing one of the aspects of acting, makes the following observation: "If an actor mimes playing the piano and the sounds of a piano are actually produced behind the scenes, there is no reason to interpret this as an action made up of two distinct elements"[14].We shall interpret the same differently: indeed, the action described is not made up of distinct elements, but that is true only behind the fifth wall (provided the piano-playing is recognized as such by the figures); however, on "our", phenomenal side of the fifth wall, the stage action is in fact made up of two distinct elements, and in this way creates meanings different from a situation, in which the actor plays "normally". In many cases, however, piano-playing on the stage is not recognized as such beyond the fifth wall. For instance, in Luk Perceval's *Othello*, one of the most important actors on the stage is the piano-player, but none of the stage figures notices his presence or the music, or the presence of the two grand-pianos that are set one (black) on top of the other (white).

Thus, let me repeat, the meaning of acting rests not only in the creation (describing) of a fictional figure alone, but in the ability to create a juxtaposition between the figure which is being created in the mind of the spectator and the actor impersonating it in his/her phenomenological appearance. This enables the process of simulation to take place. Of course, as mentioned above, the actor does not have to play a human being inhabiting the fictional realm: he/she may equally well play an animal, or an object, or an allegorical figure. But what counts is the relationship between the fictional animal, object, or allegory and the physicality of the actor playing it, his/her costume, make-up, etc. This is how the aesthetic function surfaces. We must not forget, as often happens in recent theoretical and ideological writings, that theatre is an artistic medium, and as such reveals features not to be found in non-artistic texts. What counts is *how* the meaning is created, and not necessarily what the meaning is in the referential sense. What also counts is *how* that means to the spectator. We do not have to agree with the author and/or director or with his/her ideology, but we can recognize and appreciate the artistry of the work. And vice versa, even when the ideology fully complies with our worldview, we do not have to admire the often doubtful artistry of didactic or propagandist texts.

The relationship between the actors' bodies and the denoted fiction may even be more complicated than one can imagine. For instance, in Tristan Tzara's *Heart Made of Gas* (*Le coeur a gaz*) (1920), the figures appearing on the stage, represent different parts of a man's face, his eyes, mouth, brow, forehead, chin, etc., all engaged in a dialogue, and each ostensively stressing its own *deixis*. It is obvious that their use of the language does not signify the use of any language of their fictional counterparts. As an utterance, language exists on one side of the fifth wall, the phenomenal. In the denoted fictional world, the separated parts merge, forming a human face, hence a metonymy of the whole body and personality. The humour of the play derives from the fact that different parts of a face are enacted by live human beings and are each animated and supplied with attributes of a live human being, equipped with ability to express itself. Since this feature applies to all the metonymic "representations" of a single (?) human face, we have an uncommon situation, in which several metonyms appear simultaneously on the stage, each denoting part of the same signified. Thus, the whole (the actor) represents, say, the nose, being a metonym of a face, being a metonym of a human body. Thus, in a rare example, the whole represents a part of the whole. *Totum pro parte* represents *pars pro toto*. And that unusual signification is the source of humour.

At any rate, the actor is revealed and conceived as the signalling – as well as metonymic – cause of the image of the figure in the mind of the spectator, while at the same time, paradoxically, in reverse, as the immediate effect of activities and mental and emotional processes (of this seemingly the same figure)[15] that are not objectively visible and audible on the stage. This establishes a cause-and-effect chain, which is governed by logic and establishes verisimilitude of temporal and spatial contiguity: owing to the actors' appropriate behaviour and utterances, their cause, although fictional and invisible, becomes as it were more real. Here metonymy undergoes transformation into a whole that is synthesized by the spectator. It becomes a conceptual blend, which, incorporated into the space of memory, is constantly confronted with the changing reality of the stage. The one influences the other in the continuity of cognitive reactions. This is how the chemistry of the theatre operates. The spectators for their part understand that the actor's utterances and behaviour result from a sort of contract, the initial agreement, that they result from a situation that is as if the actor were someone else, or for the duration of the performance had the ability to fill his body and mind with someone else. The actor simply changes his/her deictic costume. Hence, the actor's reactions to the outer world mean reactions to the world different from what we see and hear on the stage. Acting is entirely conventional, and one has to accept the convention to avoid the sense that what is enacted on the stage is ridiculous or abnormal. And one more thing: as observed earlier in this book, in reality the future does not exist in any material sense, and it is not foreseeable in its entire complexity and detail; it remains a possibility, determined by events and actions occurring in present or already passed. In the theatre, the situation is strikingly different, because the future is given and preordained. Still, we see the actors engaged in sorts of actions and endeavours, as if the future was not given, and they had power to shape it. This brings the unfolding fictional future even closer to the one we perceive as belonging to our own world.

The actor signals all this by really carrying out his/her profession on the stage, while at the same time he/she usually – depending on the style of a given production – signals that this is a sign of the behaviour that is natural in the world inhabited by the figure he enacts (describes). As I have pointed out above, acting in theatre is at least partly oriented towards itself, i.e. the actor wants to show both the effect of his/her endeavours and also the creative process itself (in film, at least in the recent style of acting, the actor concentrates on the effect only).[16] But we know that everything that results from this actorial "pretending" of something in him/her is really in us, in our spectators' minds. We see a live person who in his/her general features is or may be similar to the enacted live person in the fictional world. This is additionally corroborated by the fact that other figures inhabiting that world take this same person as one of their kind, a real human being, and not as an actor enacting a figure. This of course is conveyed through the signals generated by other actors on the stage, which enable non-mimetic modes of acting to be introduced that are not psychologically "convincing" and are not verisimilar to the "natural" ways of behaviour, dressing or articulation. Even though the behaviour and "acting" are conspicuously "out of character", other fictional figures read them as natural. This is the case in particular with opera, mime and various streams of a-psychological theatre (Brechtian, avant-garde, post-dramatic). One does not really expect realistic acting from an opera singer in a death scene, for it is clearly not possible to enact dying and sing at the same time. However, in dramatic theatre, one should not overstep "the modesty of nature" as Shakespeare put it: wrong casting, for instance may result in the destruction of the deictic costume, and even though the well-trained actor does everything he/she can, the fictional present time does not appear. Why not? Well, as was the case with Aquilla Theatre's experiment with *Romeo and Juliet* (2006), a bearded Juliet constantly reminded the audience that who they are watching is a male actor, and that made it impossible for him to establish an indexical relationship with the fictional Juliet. The beard anchored the actor in the real time and space, and sex, making it impossible for him to make any use of the new deictic costume. In other words, the actor remained the only source and cause of his utterances and actions, unable to present himself as the effect of events occurring beyond the fifth wall.

This explains why acting in theatre is so different from that in films: in theatre the actor aims not only to create a figure, but also to show himself in action, in the creative process of building or describing a fictional being, whereas in film the orientation is to hide the process and concentrate on the effect only. This is particularly visible in the contrast between the two mediums, when films are screened on the stage as part of the theatrical performance (e.g. The Wooster Group's *Hamlet*).[17] The goal of acting therefore is not only to create an image of another human being in the spectator's mind, but also to present the methods and techniques of the act of creation. Thus, let me stress this once again, the true "meaning" of acting is not only referential but also aesthetic, a combination of the two. Again, as we can see, in theatre the relationship of the signified to the signifier is of utmost importance. What counts is *how* meaning is created, and not only what the reference is. This is why I cannot agree with J. Veltruský that different signs in theatre can be synonymous: indeed, referential synonymity may be achieved beyond the fifth wall, but the relationship of the signified to the signifier will always be different.[18] For similar reasons I cannot agree entirely with Eli Rozik who insists that acting refers predominantly to the fictional figure, and that the actor, while deflecting the reference, does not engage in promoting a self-image.[19] From what I have been trying to show so far in this book, he/she is engaged in both, because the creation (description) of the fictional entity is not the aim in itself: as I have argued, it is the relationship between the signified and the signifier that generates meaning in theatre, and although the referent might be the same (say, King Lear), it is not the creation of a single actor only; it is created also by the design of the costume, make-up, music and by other actors/figures. As is the case with master musicians, what counts is not only what they play, but how they do it and how it affects the recipient.

The body theatrical

Let us imagine that a male actor is impersonating a woman figure by sig-nal-ling a code of behaviour and costume recognized in a given culture as feminine, even though the fact of acting and impersonation is not recog-nized in the fictional world. There, beyond the fifth wall, the person is a woman, and not a man acting as a woman. Thus the iconic function of the real body of the actor is different in each of the semiotic orders involved, on both sides of the fifth wall. And yet it is a link that blends both of the spheres and their contexts. The body continually signals spatial and causal contiguity with the created world, and thus activates the indexical relation-ship, which is the key factor in evoking fiction. The actor might wish to hide his own male attributes, making them invisible or transparent, as was the case in Elizabethan theatre. In this case, the complex sign created on the stage will be subordinated to the order of the fictional world. The actor is trying to hide the iconic attributes of his body (through the costume, make-up, wig, gestures and movement, voice pitch), although their existence is usually visible and known to the spectators. In other words, he wants to be perceived as a woman, and not as a man playing a woman. However, he might also try to foreground his male attributes, through which he fore-grounds the domination of the phenomenological order of the stage. Each of the decisions behind the actor's art will determine its stylistic affiliation, will deepen or weaken conventionality, and may articulate ideological credo, or bias. Moreover, this will have a considerable impact on the reading of the whole production. In Maja Kleczewska's *Macbeth* (2006), the Weird Sisters are presented as drag queens, who are constantly present on the stage, and that alone provides an aesthetic and ideological context through which the whole has to be perceived. Also, we might have a situation, in which a male actor impersonating a fictional female is recognized as such on both sides of the fifth wall, as is the case in theatre-within-the-theatre convention, or in Edward Hall's *Merchant of Venice* (Propeller, 2008).

Thus, the effect of theatricality also includes the actor. We see both, the actor and the figure, even though the latter is seen in our mind's eye only.

Consequently, the chemistry of the theatre includes also the actor, the figure and the spectator, who, with their retorts, are brought together in the unique reaction which is theatre communication. The fact that we "see" the figure does not result from any sudden surrender to visual hallucinations that make us stop seeing the actor and discern someone who has unexpectedly materialized on the stage; but, as I have mentioned, our mental image or blend of the figure, created as we watch, merges, if not wholly, then to some extent, with the physicality of the actor creating it. The blending of the two marks the unique artistry of theatre and acting, although it may also result from the spectator's individual attitude. A kind of feedback or closed circuit follows, by virtue of which the actor and the figure "live" in the same time, and because of this in the physical and biological (bodily) sense they cannot "live" apart, next to each other. A physical and ontological illusion of a union of fiction with reality takes place. We stop noticing the actor, for he/she becomes "transparent"; it is not he/she who says "I" of him/herself, but the figure. When stage signs become transparent, it means that we are adapting the figures' model of perception (similarly, in the process of reading we stop noticing the print). We hear the words of the figure, not the actor, for the figure "speaks" for itself; of course we shall never actually see the figure speaking on the stage, but under illusion we may treat the actor as the figure. This opens the spectator's access to the feelings and emotions of a fictional being, and this may take the form of a parallel experience. Thus, experiencing fiction becomes a paradoxical reality, and the fictional being becomes a live human whom we see on the stage.

So, as an icon, the actor contributes significantly to the overall features and characteristics of the imaginary construct we somewhat misleadingly call a "character". While watching the performance, we are also led to understand that events and utterances occurring in the fictional realm have an impact on the actor's behaviour: he/she responds to questions asked, takes part in dialogues, cries, rages with fury, runs away from danger, or even dies from wounds suffered in fictional battle or commits suicide in fear of humiliation. And all for Hecuba, as Hamlet would say. What's Hecuba to him or he to Hecuba? The answer is obvious: nothing. However, in the creative activity we call acting, this is essential, for an illusion is created as if all affairs and matters connected with Hecuba and her world had a direct effect on the actor. Crying for Hecuba is the effect of what is happening in ancient Troy, but also the effect of an actor doing his job, his technical skills. Also, the actor seems to posses the ability to shape the fictional future, even though it is given. Two streams of time convey two sequences of cause and effect. And it does not really matter whether or not the actor identifies him-/herself with Hecuba.[20] Concomitant to his/her verbal reactions to the imagined fictional realm are the non-verbal ones: gestures, facial expression, tears, anger and the like, all of which heighten the emotive function, or modality. The spectator may experience the fictional being's emotions and feelings.

In other words, acting is essential to both the establishing and the crossing over of the ontological borders between the two worlds, the fifth wall if you will. And that fifth wall is essential to the meaning-generating process of the performance. It is therefore the actor who signals throughout his/her performance that whatever he/she does or says, whatever happens to him/her, is the evidence of his/her ability to transgress and cross the fifth wall, and he/she describes the fictional world and presents him/herself as the effect and "eye-witness" of whatever happens there. Consequently, an illusion is created not only of the present time shared by the actors/figures and the spectators, but also of their shared future. As I have been saying, a kind of feedback system or loop is created, and the mental record registered in the mind of the spectator, his/her memory, is paradoxically "played back" on the stage, while at the same time what is apparently replayed, that is the actor and his/her acting, has a reverse effect on the shape of the record (for of course in the mind of the spectator the structure of the figure develops in time in a sequence of conceptual blendings). The art of the actor reveals the mechanism and rules of his/her activity, while the experience of the spectator that accompanies this is a confirmation of the revelation of the aesthetic function of acting. However, we cannot always rationalize this experience. But it seems that when the imaginary structure dominates, it causes the actor to be perceived in greater degree as the "result" of the figures' actions, mental processes and emotions, which can lead to the identification of the figure with the person on the stage (and in turn of the spectator with the figure), whereas when the actor dominates, he/she causes us to see him/her as the mainspring of what we are watching and then we may admire his acting skill, professionalism, etc. or, on the contrary, condemn their absence. It may be said that both these ways (only apparently negating each other) of perceiving the actor are characteristic of the theatre. What is more, individual styles of acting can be described from the point of view of the domination of one or the other tendency.

The essence of an actor's art, then (in live theatre) is the ability to create a "closed circuit" or feedback system in the relations between the actor, the figure signalled, the spectator (with the figure imagined in constantly evolving shapes: the figure imagined is a process, not a fixed entity), and back – the figure imagined, the actor and so on. We, the spectators, take the input spaces of the actor and the figure signalled, and blend them in cognitive responses, creating an imagined figure, which may be treated as the resultant structure of the process. This is then related back to the stage and, again, back to the spectator, and so on, in a communication continuum. The process is possible because of the actor's ability to create fictional time structures. Without him/her no theatre is possible. Needless to say, it is also impossible without the spectator. And it is not enough to say that the art of acting is established by an indexical relationship of the actor and the figure. The same occurs in the case of children playing, street parades or fancy dress balls. It is the fictional present time and the perceiving mind of the spectator that transforms playing or games into theatre. Thus acting may be seen as embodied time.

The estrangement effect

The inadequacy of the language used by the actors to the phenomenology of the stage (or the other way round) will appear in a different and often striking variant in the case of the so-called estrangement or distancing effect.[21] This style of acting has been dealt with in numerous books and articles, so I shall give just one conspicuous example. In Roberto Ciulli's recent production of *King Lear* (Theater an der Ruhr, Mülheim, Germany; seen at the XI International Shakespeare Festival in Gdańsk, Teatr Wybrzeże, August 2007),[22] the ambiguities of language are deepened by the fact that each of the actors performing on stage impersonates at least two distinct fictional figures. One of these figures is not Shakespearean at all, and for that reason, and also because only Shakespeare's text is used throughout, what is true for one of the figures in its specific context is not necessarily true for the other. Moreover, whatever the situation might be, the relevance of Shakespeare's language to one of the figures may always be related to the implied relevance of the same utterance to the other figure, by which the network of possible relationships is deepened and becomes many-layered. This means that all the verbal utterances, constituting the stage-speech, and including speech-acts, appear simultaneously in distinct contexts, by which they generate different meanings depending on their relationship to one of the two fictional speakers (played, as I said, by one actor) and their deictic axis. Thus the meaning of stage-speech may be drawn from the relationship of the referential world, as created verbally by live actors, to one or the other of the two fictional figures to whom a given speech is attributed. The described situation, rather rare in theatre, relies on the actor's ability to create more than one assumed *deixis* of fictional figures. On top of one deictic costume, the actor puts on another one, which does not cover the former one completely.

In Ciulli's *King Lear* all the actors are seemingly blind, from the very start of the performance. More precisely: through the acting technique adopted, they indicate that they cannot see. But they do this unobtrusively and very subtly, so subtly indeed as for it to pass unnoticed by many of the audience and even the critics. They move only reluctantly, and if at all, then only uncertainly. Their eyes are wide open and they try not to blink, by this means conveying the impression that their eyes are motionless and thus do not react to the external world. Additionally, the actor "playing" Gloucester hides his eyes behind extremely dark glasses. Naturally, since the figures in Shakespeare's play are not blind, the spectator requires an explanation for their blindness in this production. In the beginning, however, only one thing is certain, namely, that the fictional "actors" are blind, that they are cut off from the external space and the geometry of the world around them, which has a decisive impact not only on their behaviour, but also on the proxemic (and metaphorical!) relationships between the denoted fictional figures. Cut off from spatial relationships, external to the perceiving mind (as Kant and others have shown us), all they are left with is time, which is an internal quality of the mind. In this way, the production concentrates on temporal issues, which may be related both to epistemology, the cognition of reality by humans, and to rules that govern the theatre as art, of which those concerning the time structures involved are of uttermost importance.

All the actors taking part in the performance are on stage (which also means a stage in the fictional realm) all the time; there are no entrances or exits. They stand still, waiting for their cues, and when they speak, they address the space, the darkness, that surrounds them. Some take a few steps, feeling the furniture, the large, old-fashioned tape-recorder that stands on the table in the centre of the stage and plays an important role in the production, or the bodies of others, looking for support from them (Figure 8). Only Edgar, half-naked and filthy from the start of the performance, seems to act independently of the evolution of the "action"; he crawls over the

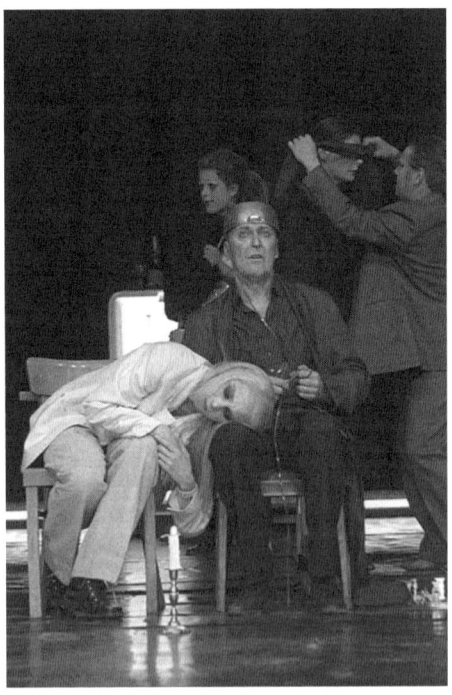

Figure 8   Lear (Volker Roos) crowned with a saucepan, holding the recording tape, with Cordelia/Fool (Simone Thoma) at his side (photo by Ryszard Pajda)

ground, as if practising animal movements long before the scene when, in the framework of the plot of Shakespeare's *Lear*, he turns from a nobleman into mad "Tom of Bedlam". In Ciulli's *King Lear* he appears as an actor practising the role of a naked beggar from the beginning. Sometimes the actors collide with a wall that is invisible to us and that demarcates the space of the stage, but this too is indicated very subtly, only as it were signalling the rules of the theatre, which enable whole worlds to be created out of practically nothing in the material sense. Paradoxically, we, the spectators, see the actors bouncing off the wall that is invisible to both the fictional figures and the audience. However, the fact that we do not see the wall, does not mean it does not exist in the fictional realm: what the figures sense with their touch (an act of ostension), we perceive through convention. We understand that the bouncing off signals an indexical relationship, based on spatial contiguity, between the fictional figure and the wall. This means that the fictional space is an enclosed one, similar in dimensions to the theatre stage, enclosed by solid walls; this also means that all the actors are set within that space (which of course contradicts the multitude of changing spaces employed by Shakespeare in his play). But this does not mean that they are on the stage of Teatr Wybrzeże, where there are no solid walls enclosing the platform on all sides. In this way a fictional space is denoted through actors' gestures and movements.

At all events, we understand that the actors, playing blind actors, are creating a fiction. The space created is also a fiction, because the rehearsal is taking place not on the stage of Teatr Wybrzeże, but in some room, in some space built metonymically, and so conventionally, theatrically, without the participation of the audience (whose presence is not signalled by the actors). Thus its time, too, is different from that of the real performance. We are therefore dealing with a typically theatrical gap between the time and place of the audience and the time and place of the created world. But – as I have said – in the production under discussion there is a duplication of fictional realities, which is achieved basically through the multiplication of *deixis* of denoted fictional figures. At the same time we see that the departure from psychology does not involve a simultaneous erasing of the borders of theatre or an undermining of its rules. These are maintained throughout the whole duration of the performance, which is modelled in such a way that metatheatrical impulses take on significance: the spectator is constantly reminded that what he or she is dealing with here is theatre, though not the kind of theatre that strives to create an illusion of anything. This theatre does not pretend, and the fiction that it creates is self-referential in a high degree. We watch a theatrical performance that is stylized as a rehearsal. But this is achieved by the use of every available theatrical resource. Purified of the illusion of the world outside the theatre, of the psychology of traditional productions of Shakespeare, it delves deeper in its aesthetics, revealing and explaining the rules that make it possible for the theatrical message to take shape in the form conferred on it. This creates a theatrical metaphor of human life and of the stage, of life and acting, playing roles. It is not a new metaphor, indeed it is perhaps one that is over-exploited today, but in theatre (as in all art) what counts is not "what", but above all "how".

In spatial terms, the blind actors perform, as it were, each alone and for him/herself. They utter their lines into space, which for them is a scalar, often without direction or orientation of any kind. They also do not create by ostensive means the fictional times and spaces that belong to the world of Lear; the latter is constructed verbally, with a marked distance between what is being said and the world that the actors are set in. They do not pretend that they are in pre-Christian Britain, they do not pretend to be Lear, Goneril or the Fool. They inhabit a modern world, with refrigerators, electricity, tape-recorders, etc. In this way a more complex relationship is established, and for example, the actress becomes related not only to Cordelia and the Fool she is playing (or rather, uttering their verbal utterances), but also to another fictional figure who is not an inhabitant of the world created by Shakespeare and "lives" at a different time and in another space. Also, to make matters even more complicated, the three fictional figures enter into all sorts of mutual relationships.

The words uttered by the blind actor indicate that the denoted fictional figures in *Lear* are not blind. Consequently, their actions cannot be the same. You cannot blind a blind person, if we take Gloucester as an example; you cannot write and read letters. This creates a certain confusion in the spectators for it becomes noticeable that what we observe happening on the stage is not in the time and space indicated by Shakespeare's text, and the actors on the stage do not fully comply with what traditionally has been labelled as Shakespearean. And yet the action still takes place in a fictional space and a fictional time, not in the here and now of the audience, even though the effect of the fourth wall is somewhat weakened by the fact that the actors do not have to pretend not to see the audience; since they are blind, or present themselves as such, they *really* cannot see the spectators (of course, their orientation in space is to some degree possible, owing to other senses, such as smell, hearing or touch). But, as I have indicated, they only pretend to be blind and the last scene, when some of them are "relieved" of their impediment, may stand as a proof; they seem to enact blind actors rather than the fictional figures from Shakespeare's *Lear*. This preserves the rudimentary level of fictionality, necessary for theatre to create meanings according to its specific rules, but also foregrounds the estrangement effect in the style of acting.

As a consequence of their loss in space, it is rare for the actors in Ciulli's production to succeed in saying anything in "face to face" dialogue. They are cut off from the geometry of the world, which gives us orientation in space. Their words cut across the space in a variety of directions, often independent of the location of the interlocutor. This may of course be interpreted metaphorically as an attempt to emphasize their isolation, their imprisonment within themselves, lack of mutual understanding and inability to communicate with one another, a feature that may also be applied to Shakespeare's figures (fictional layer 1 enters into all sorts of relationships with layer 2).[23] When they sit opposite one another at the table, they usually seem to conduct their dialogues not with another person in their shared present time, but with the voice rendered by the tape recorder: with past time. For the conversation with the recorded voice is a clash between two times and two ontologies. By this means, the past marks its permanent presence in the implied consciousness of the stage figures. Moreover, from the point of view of a blind person who has lost a sense of space, the human voice is the only attribute of a human being (as in a radio play): it does not really matter whether it is "live" or recorded. Thus playing the voice from the past brings it into the non-spatial present. This also invites further metaphorical interpretations, for the tape-recorder may be seen as a metaphor of memory (a human mind incapable of perceiving objective space, cut off from the geometry of the world, relies on time only). In mimetic terms, the recordings may result from previous rehearsals.

The proxemic relations, the blocking, and also the failure of eye-contact and of meeting in time contradict the principles of inter-personal communication as we experience them in daily life. Hence the distancing and far-reaching "de-psychologization" of this production, which Hans-Thier Lehmann considers one of the features of the postdramatic theatre. Naturally, "de-psychologization" involves one of the fictional layers created here through the relationship of the Shakespearean figures to the actors. The actors do not "represent" anyone apart from themselves: they do not play figures from Shakespeare's drama, but figures from Ciulli's production. Thus real actors impersonate fictional actors. They only speak the Shakespearean text, while remaining themselves. This is achieved through the lack of congruity between the verbal component and the acting. For instance, the voices of the actors are not modulated theatrically to "suit" the figure from Shakespeare's play; their harshness sometimes grates on the ear (Cordelia). For a considerable part of the performance, Edgar emits only animal-like, inarticulate sounds, howls and mumbles; he is really mad, or – as we understand after his "transformation into a human being" – he does not believe that in his situation words or articulate language could change anything. He does not believe in the possibility of communication, in dialogue (but he too turns on a piece of text from the tape-recorder: so he understands human speech). Again, in mimetic terms, Edgar-the-actor may be preparing for his role, producing inarticulate sounds. However, even though the actors do not impersonate Shakespeare's figures in any psychologically convincing way (to use a dated criterion), they do signal their attitude to Shakespeare's text, even in those cases when it is based on indifference or total lack of congruity. But that is also meaningful.

All of this creates a dialogic situation between Shakespeare's text and Ciulli's production. The director does not so much "interpret" the play as insert it into a production of his own creation, which in many ways is not predicted or implied by the play. At best, he translates the verbal into the visual, creating a transmutation rather than an adaptation.[24] This creates yet another level of meaning: Ciulli consciously plays on the relationship between the fictional world as indicated by Shakespeare's text (along with its traditional readings) and the fictional world created by what we see and hear on the stage. The former is basically verbal and supported by congruous acting, while the latter is non-verbal in the sense that, as I have indicated, the language uttered is not part of the world inhabited by the actors. It is external, if not alien. This creates a paradoxical situation, by which the realm denoted by language is not the world in which the actors pretend to live. And yet they signal to us that their world is not the reality of the spectators. The effect of this situation is that we observe a split of *deixis*, or, rather, its multiplication. The "I" of the actor we see is not the "I" of the real person whose profession is acting and who appears on the cast-list; naturally, this would have been a normal theatrical situation if the actors had implied, through acting, that they had assumed another "I" to whom the verbal utterances "belong". But they simply do not act in that way. As the immediate result, we come to understand that the language belongs to someone else (a third "I" and *deixis*) who is predominantly implied by Shakespeare's text, and not by the actor's attempt to assume the role. This leads to the rise of two fictional levels, which are in constant play in Ciulli's production. The result is that every verbal utterance refers to three distinct worlds: the real world of the stage in Teatr Wybrzeże, the fictional world of the actors rehearsing *King Lear*, and the implied fictional world of Shakespeare's play. The side effect is that the verbal component becomes disintegrated or detached from stage speech, and through this draws the spectator's attention towards itself and its material context, consequently creating all sorts of unexpected equivalences and relationships. The example of Ciulli's production shows clearly that acting does not have to rest on impersonating "psychologically convincing characters".

Part III
Retorts of Time: Temporal Conventions

8
Soliloquies and Asides: Who's There?

The strange case of speaking to oneself

When discussing theatre conventions, I am using mostly Shakespeare and his contemporaries as the major dramatic source, partly because their plays are widely known and read at all levels of educational systems in many countries, but predominantly because those playwrights were conscious commentators on their own art, and the "chemistry" of their plays very often reveals the rules by which particular elements combine on the stage, and by which they act in different conditions. In many ways in *Midsummer Night's Dream*, the scenes involving the Athenian craftsmen may be seen as an artistic equivalent of the theory of theatre art. The Elizabethan and Jacobean plays not only present all sorts of "stories" or fables, but to a large extent "talk" about themselves, and explain to us the rules of theatre "language".

Shakespeare's soliloquies and asides are so numerous and well known that it is absolutely impossible to discuss them individually in a relatively short chapter.[1] I do not think a Shakespearean play may be found without at least a single soliloquy or an aside.[2] Also, in today's theatre, in comedies and farces in particular, we find a great abundance of various uses of the convention. Additionally, what is tempting here for a theorist, is the fact that when dealing with any subtle subtlety of theatre, one inevitably has to deal with basic problems and definitions of the art.[3] Nevertheless, I hope that through the presented analysis not only a convincing definition of the soliloquy and the aside will be shown, but a verifiable conclusion may be reached, which will explain some of the ways in which theatre as art has the potential to play with meanings and conventions contained in the dramatic text. I shall therefore concentrate on the possibilities of staging alternatives, which show that the decisions undertaken by theatre directors may actually change generic markers of the dramatic text and also change the signalled stage conventions to the extent that a given passage, which in the written text is, say, a part of a dialogue may evolve into a soliloquy or an aside, depending on the preferences of the director. This inevitably changes the meaning of the passage and may also become an additional source of humour and so on. It is therefore safer to talk about certain tendencies or orientations rather than strict demarcations between particular modes of stage utterances. A soliloquy may reveal dialogic features, and may even include occasional asides, expressed directly to the audiences; and *vice versa*, what seems to be an aside, uttered during ensuing dialogue, may reveal the features of a soliloquy. Some scholars add an interior monologue as yet another distinctive mode of utterance in drama.[4] Thus, what was originally intended to be a soliloquy by the playwright may, in stage practice, lose its specific features and turn out to be a monologue, or a text read out from a book or letter; an aside may appear in places not foreseen by the author, and *vice versa*, what originally was intended to be an aside may lose its distinctive features. Moreover, speech attributions may be altered, as in a recent production of *Hamlet* by Monika Pęcikiewicz (Teatr Polski, Wrocław, 2008), where it is Ophelia who delivers (for reasons unknown) the "To be or not to be" soliloquy.

Let us observe first that both the soliloquy and the aside are not modes of speech selected out of many possible forms of verbal expression to fulfil specific needs of the figure; on the contrary, they fulfil the needs of the author (playwright and/or director) who makes the selection in order to convey information that otherwise could not be transmitted or would demand much more time than the strict stage time economy would suffer. That means that both are conventions and draw our attention to their conventionality because they are conspicuously different from other modes of speech, such as dialogue, and are not a mimetic imitation of an ordinary human verbal behaviour. In everyday experience we usually do not talk to ourselves (in fact, talking to oneself out loud in public is usually considered a sign of insanity) and do not make verbal asides that our interlocutors cannot hear, not to mention the fact that the addressee of asides is non-existent outside the theatre. It is reasonable therefore to treat both types of utterances as not occurring as a normal verbal behaviour in the fictional world, but as signs of the state of mind, emotions and feelings of the figures and also as signs of authorial "intervention", the intention of which is to provide the spectator with information that is necessary to increase his/her awareness and understanding of the fictional world. It does not mean that the original stage convention could not have been different, and could in fact be a sign of someone talking out loud, so that eavesdropping was possible.[5] However, one of the most important functions of soliloquies is to differentiate the levels of awareness between the stage figures and the spectators, a strategy vital for creating suspense, mystery, for providing insights into a figure's psyche, and so on. The differentiation is possible only when the information provided by the speaker is not known to other figures, and the spectators become intimate witnesses to all sorts of inner confessions. Moreover, the fact that nobody else can hear the soliloquy is a proof of the speech's truthfulness (usually one does not lie to oneself); if on the other hand we treat a soliloquy as an actual speech act, the speaker would inevitably be conscious of the risk of being overheard or someone eavesdropping, which would naturally have an impact on what he or she says.[6] This is one of the reasons why James Hirsh's claim that all soliloquies in Renaissance drama are signs of actual speech is not entirely convincing, and in many cases it is contradicted by stage action and immediate verbal context, as some of the examples provided in this chapter show.

It will therefore be safer to say that a soliloquy is a convention, by which a speech is uttered without a clearly marked addressee other than the speaker, whether or not it is uttered aloud on the other side of the fifth wall.[7] Therefore, if in a given production Hamlet is aware of Polonius's and Claudius's presence during his "To be or not be" speech, it may lose its features of a soliloquy and gain the features of a monologue that has a clearly defined listener. If, however, the Prince is not aware of their presence, the speech remains a soliloquy, even though eavesdropping actually takes place on the stage. In such a case, we have at least three dichotomous options: one is that the eavesdroppers cannot hear anything, for the simple reason that a given utterance is a sign of mental processes rather than actual speech; two, even if they can hear the figure speaking aloud to him/herself, the speech, as a linguistic construct, does not contain any features of an utterance that has an addressee who is eavesdropping; and three, the speech is constructed in such a way that we understand that the speaker is aware of someone eavesdropping and adjusts his/her speech accordingly.

Linguists have drawn our attention to the various tactics employed by speakers depending on the recognition of various addressees. For instance, due to the presence of some unwanted bystanders, the speech may be communicated in such a way that bystanders will not be able to understand the full context of the utterance (in this way the speaker may select between various "types" of bystanders, some of whom may be allowed to understand the hidden meaning, as in the scene between Hamlet, Horatio and Osric). A different approach will be taken when the speaker is aware of the presence of eavesdroppers; when planning the utterance, he/she will assume a specific strategy: indifference, disclosure, concealment or disguise.[8] So, if, for instance, in a given production Hamlet is made aware of someone's presence during any of his soliloquies, we may interpret it as shaped in such a way as to deceive the overhearers. If evidence for this strategy was found in the linguistic shape of the utterance, it would prove beyond doubt that it was intended to be spoken out aloud and with a specific addressee in mind. However, in most cases (in Shakespeare) the evidence is not unanimous, which leaves the doors open to various possibilities in actual stagings of plays.

In order to prove my point and illustrate the phenomena mentioned, I shall give the well-known example from Shakespeare's *As You Like It*, where in act 2 scene 7 Orlando exits in order to lead in Adam; in this time Jaques delivers the famous monologue "All the world's a stage", in total thirty-three lines, so about two minutes of scenic time. As soon as he has finished,

Orlando returns – he has found Adam and brought some food. It is hard to say how much time this would have taken in the empirical world – say, an hour. Since no one within the world presented defines the temporal value of Jaques's monologue, our first reaction might be to accept that it is close to the time of the audience, so we have here a division of fictional time into two: the scenic one and that which runs beyond the stage (one minute of the stage time is equivalent roughly to thirty minutes of the offstage time). This would mean that we are dealing with two distinct streams of time, the one operating on the stage, and the one offstage. In this particular example, the two streams form a clear contradiction.

Naturally, a convention helps us to bridge the temporal gap between the two worlds, the one on the stage and the one offstage. In the scene under discussion, we may begin to suspect that Jaques's monologue is in essence a soliloquy, a sign of thought processes rather than actual speech, which took him about an hour. In this approach the temporal contradiction is weakened if not annulled. This implies that the real time of the real actor (i.e. human being) delivering the speech on the stage flows with a different tempo than the one of the stage figures and the one offstage. However, this seems to contradict the fact that the speech is a direct response to the Duke's words, who is the first to employ a theatrical metaphor, so it appears as a part of a dialogue and on the surface there is no reason whatsoever to treat Jaques's words otherwise than a factual utterance. On the other hand, the Duke does not respond to what Jaques says, for their dialogue is interrupted by the appearance of other figures. This leaves the director with at least two options: he or she may have the Duke listen to what Jaques is saying, and through this (gaze, gestures, facial expression, etc.) signal the monologic nature of the utterance, or present the Duke totally inattentive to Jaques's words, which would imply that we are dealing here with a soliloquy (a mental process which takes an hour or so, marking a conspicuous difference between the fictional time offstage and the real time of actor–human being); it is also possible to utter the speech direct to the audiences as a sort of lengthy aside, and that could be supported by a changed style of Jaques's acting which would signal and establish phatic function of communication between the figure and the spectators (marking a total coalescence of the fictional time of the figure and the real time of the actor). The example shows clearly that a director may change a monologue into a soliloquy or an aside, thus changing the meaning of the stage action and characterization. A soliloquy makes Jaques more isolated from other figures on stage, a person who does not share his feelings, emotions and thoughts with the others; a monologue, which often is part of a dialogue, marks his will and ability to share his inner life with the others and also, other people's interest in Jacques and his thoughts; an aside, in turn, would equip Jacques with features that other figures do not possess – with his ability to step out, so to say, of his role and address the audience directly, as if the actor and the figure were one and the same ontological and biological being.

A hole in the wall

The distinction between a soliloquy and an aside is not an easy one: for instance, Patrice Pavis in his influential *Dictionnaire du théâtre* defines both in almost exactly the same way, as an address of the speaker towards him/herself (as opposed to the usual addresses towards other figures on the stage), which results in the factual address towards the audience (direct or indirect). However, using solely that feature, we are unable to distinguish the stylistic devices from each other. Also the definition does not include non-verbal asides, which in theatre play an important role. Neither does Pavis distinguish asides made to other figures. In this chapter, I shall propose a different approach, and try to define the soliloquy as a conventional stage device, in which the speaking figure does not "step out of itself" but addresses (verbally) the imagined audience (or him/herself),[9] directly or indirectly, remaining Macbeth, Hamlet or Lear. This also means that the fictional time of the fictional figure keeps flowing with a different tempo from the time of the real actor (human being), as is the case with most stage action and utterances. And it is not the time of the audience: a short soliloquy may in fact indicate the passage of, say, several hours. Only the present time is common to all on both sides of the fifth wall. It seems therefore that in interpreting soliloquies we should take time as the key factor upon deciding whether a given utterance is spoken out loud or whether it is a sign of mental processes. Consequently, instead of guessing, we shall obtain more solid grounds in classifying a given soliloquy: if the signalled fictional time is relatively the same as needed for a given speech to be uttered in "real life", then we may (but do not have to) treat the soliloquy as a sign of actual speech; if the signalled fictional time is conspicuously different from the time needed for the utterance (as in the final soliloquy by Marlowe's Faustus), then we should treat it as a sign of mental processes rather than actual speech. Occasionally, the two types may overlap, because a soliloquy may also be uttered within a scene of dialogue: when the interlocutor keeps talking by just moving his/her lips, and we cannot hear anything he/she is saying, and at the same time another figure delivers a soliloquy, we understand that he/she has stopped listening to what the other figure is saying and what we hear are his/her inner thoughts. The latter is corroborated by the fact that the interlocutor does not hear anything that the figure is saying, but not because he/she does not want to, but simply because they cannot. This in fact is a cinematic technique, less conventional, with a greater amount of verisimilitude and an element of subjective perception, only rarely signalled on the stage.

Also, it seems important to distinguish asides addressed by the stage figure to him/herself (which do not have to be signs of real speech, but are rather the signs of thoughts), from those addressed to other figures and those clearly addressed to the audience. The former type may be treated as a variant form of a soliloquy (being a sign of a mental process, and not actual speech), the latter is a true aside, for it is a verbal address uttered directly to the audience, and for that reason seemingly violating basic conventions by which the stage world is separated from the world of the spectators. However, since the speaker seems to notice the presence of the audience, whom he/she addresses, it follows that he/she has to be set in the same time and space as the spectators. This seems to violate the basic rule of the theatre, so much dealt with in this book, the rule of accessibility. Thus, in soliloquies, the two distinct streams of time, that of the figure and that of the actor (and audience) will not change and will frequently be marked and foregrounded, whereas in asides addressed to the audience they will take a rapid turn and seem to conspicuously and thoroughly coalesce (for its duration the present time of the figure will flow with the same tempo as the present time of the actor and spectators). Addresses to other figures ("aside to X"), on the other hand, are a form of a dialogue,[10] and share all the features of (whispered) dialogic exchanges, with the condition being that at least one of the other fictional figures on the stage cannot hear what is being said. The latter situation is conventional because the laws of acoustics are violated, and we have to take for granted that the asides are whispered in the fictional world in such a way that the message is heard by selected figures only.[11]

In the case of the aside, especially the one made to the audience, the general situation of the speaker is strikingly different from that in soliloquies: the address is made in such a way that the speaking "I" of the speaker now seems different, as if he/she stepped out of the *deixis* of the fictional figure, and as if it were the real actor who *is* the figure who was uttering the words directly to the real audience, trying at the same time, through various gestures, gaze and facial expression, to establish phatic function with members of the audience. Until the moment of the appearance of this mode of speech, the actor has been pretending not to see the audience and not noticing the theatrical situation at all. Now, all of the sudden, he/she starts to notice the presence of the spectators, without the fact being noticed by other stage figures. This means that in an instant, he/she – if still a figure – has found him/herself in the seeming reality of the audience, seemingly sharing with it time, space and awareness of the theatricality of what is presented on stage. If not a figure any longer, he/she stops sharing time and space with the other stage figures present on stage, and seems to leave the fictional reality, which explains why they do not hear him/her say anything, and why they cannot even see him/her say something to any invisible someone (which would be the logical thing to observe). This also means that the ontology and *deixis* of the speaking actor change, for his/her awareness of time and space is strikingly different from the awareness of the hitherto enacted figure. Until now the actor has been a sign of the figure, and his/her words were merely signs of the fictional language used by the figure he/she enacted; in other words, until the aside occurs, what we were hearing were the actor's words and not the figure's. Now, during the aside, the person who we are hearing wants to be perceived as the figure speaking to us direct, at least it so seems to us. This means that the actor's stage speech becomes identical with the language of the figure. But that is totally improbable, because fictional beings cannot appear or speak in our physical reality. So, other explanations have to be found for the following problem: who exactly is the speaker, and whose words are we hearing, the actor's or the figure's or both? Who's there, behind the appearance?

However, it must be observed that the violation is the logical consequence of the gap in temporal structures, a hole in the wall, if you will: the present time of the speaker of an aside ceases to be the present time of other figures, for it has become the present time of the spectators, at least for the duration of the utterance. Especially in the case of an aside, if addressed directly to the audience,[12] which we consider to be "true aside", both the performer and the fictional figure seem to be noticing the real spectators' presence, which implies that the actor has either stopped acting and is speaking as him/herself, or – as a stage figure – has made a rapid shift from its fictional time and space and has miraculously landed in the *hic et nunc* of the audience, by which the fictional present time of the figure becomes the real one of the sides involved. The unity of time and space between the performer, the figure and the spectator may be additionally signalled by a changed style of acting, gestures, facial expressions, gaze and the like. So, during the aside, the time of the fictional figure seems to coalesce with the time of the real actor (human being). Its present becomes, or seems to become, factually congruous with the actor's present.[13] This means that the ontology of whoever utters an aside is different from the figure hitherto enacted by the same actor.

Moreover, as indicated above, the speaker of an aside implies through his/her words and gestures that he/she lives (albeit temporarily) in the same time and space with the members of the particular audience gathered on the particular night. His/her present time seems to be totally coalescent with the present time of the audience (otherwise it would be entirely impossible to establish a phatic function). And this is exactly how many members of the audience treat asides – as utterances of the real actor who is identical to the figure he/she enacts. This could imply that it is the figure who is speaking to us. However, a closer analysis shows that this is yet another theatre convention. So, the options are that, in the case of a true aside, what we are seeing and hearing is an actor only, who has for some reason abandoned his/her role, or stepped outside it (distanced him/herself from it), or, that the fictional figure has miraculously materialized in the shape of the actor. Both of the proposed solutions, however, are not satisfactory to solve the rising contradiction and improbability. The actor is not allowed to speak as him-/herself, unless spontaneously or improvising, and the miraculous "transubstantiation" of the figure on stage is not very likely. The observed phenomena last for the duration of the aside only, after which, the actor returns to his/her former style of behaviour, pretending not to notice anything outside the fictional world, returning to his former deictic axis (where "now" and "here" mean something different from what they mean to the audience). So, let me repeat the question: who actually is the speaker of the aside? I shall return to this question shortly.

The similarity of some forms of monologues or soliloquies to asides is superfluous: in most cases they are not a breach of a dialogue and a concomitant suspension of one of the streams of fictional time; when they are, they become an aside. Also, the soliloquy does not signal a shift of the speaker in time and space (it occurs during an undisturbed sequence of events and utterances). And usually, it is not an open address to the spectators. We may, of course, encounter mixed variants of longer asides or soliloquies with internal fragments of written texts that are read aloud to us. For instance, this is the case of Julia's soliloquy in *The Two Gentlemen of Verona* (1.2.102–27), in which she is trying to put together scraps of a letter that she has just torn. This also introduces a feature, frequent in soliloquies, namely, a dialogic composition.

However, there is a type of stage speech which takes the form of a soliloquy uttered in motivated situations, for instance, as August Strindberg proposed, when an actor is reading his/her part aloud, a figure is presented reading a book or a letter aloud, a servant-girl is talking to her cat, a mother is prattling to her baby or an old spinster may be chattering to her parrot. This kind of speech is a sign of someone actually talking, usually without any other persons being present, or the speaker not being aware of someone present. But the utterance is a sign of an actual speech within the fictional world, and also, it usually has an addressee who is not the audience. So, if Hamlet, instead of presenting "To be or not to be … " in the "traditional" manner, reads the text out of a book which he is holding (as if someone else wrote something suitable to his state of mind), the utterance ceases to be a soliloquy and becomes an example of reading a text aloud to oneself, a kind of hidden address *ad spectatores*. In that case, the words do not necessarily mean Hamlet's mental turmoil, the fictional time is not suspended and the oral act of reading becomes a sign of either silent or loud reading. In a production of *Hamlet*, in Gdańsk, Poland, entitled *H.* (*Teatr Wybrzeże*, 2004) and directed by Jan Klata, the text of this particular passage (i.e. "To be or not to be … ") is presented by amateur actors during a casting session that Hamlet organizes for the play-within-the-play. Similarly, a soliloquy would cease to be itself if Hamlet's delivery was heard by other figures present on stage, which could be signalled by their gestures, nodding or applause, with all responses perceived by the speaking Prince. Again, all those variants depend on the decisions of the director. Whether or not we approve of them, is another question that cannot be dealt with here at any length.

In the eavesdropping scene in *Twelfth Night*, the comments about Malvolio's soliloquy made by Sir Toby, Sir Andrew and Fabian, leave no doubt that the three conspirators hear perfectly well what Malvolio is saying, whereas the audibility of utterances does not work the other way round: Malvolio does not hear the asides made by the three men. The kind of speech delivered by Malvolio Manfred Pfister calls a "motivated soliloquy", which may be contrasted with the unmotivated or conventional one.[14] However, I think that that kind of utterance ought to be treated rather as a monologue, which may in fact often reveal features of a dialogue (since usually they are not self-addressed to the speaking person, but have an addressee, such as the baby, the cat or the parrot, to use Strindberg's examples).[15] Malvolio in the scene mentioned above seems to reveal the habit of talking aloud to himself (which in comedies and farces is not an infrequent feature of figures), but even he uses the found letter as a sort of point of reference, which motivates his (verbal) behaviour. Please note that in this particular case time flows with the same tempo as in any other verbal activity of humans; it is, for instance, defined by the verbal reactions of the other figures. In conventional soliloquies time remains usually non-defined (and therefore leaves the interpretation open to particular theatre productions). And the "asides" made by the three men are not true asides, for they are not addressed directly to the audience and each utterance made is heard by the other two men. Thus, they may be treated as dialogic exchanges, whose comic effect derives from the fact that they may be related to an utterance of a person who cannot hear them.

The question of asides that are not addressed to other figures or to the audience is problematic, because in many ways they are more like brief soliloquies, often interwoven into dialogue. Their significance rests in their ability to signal the speaker's state of mind, his/her emotions, during the ensuing dialogue. They often serve as a commentary on what we are hearing and seeing. This means that their time flows within a different stream than the dialogue proper, which explains why nobody notices time gaps caused by what sometimes is a lengthy speech; they are not signs of actual verbal utterances, but of mental processes. This has often puzzled commentators, but it seems that utterances of this kind are conventional and have to be treated as utterances outside the temporal stream of a dialogue. For instance in *King Edward III* we find an awkward scene (2.2.75 ff.), in which King Edward converses with the Prince, and produces asides that are much longer than the dialogue proper. When Lodovick and the Prince exit (in l. 109), the King continues using the same style of speech in what undoubtedly is a soliloquy (ll. 110–15), since he is alone on the stage. Does it mean that an aside has evolved into the soliloquy? It seems that we ought to distinguish more rigidly "asides" addressed to the audience from ones that are not and serve the function of a sign of mental processes rather than actual speech. Also, we should take into account the context, both verbal and non-verbal of asides and soliloquies: depending on the context the same speech may be either an aside or a soliloquy. If the speaker does not signal his/her awareness of a theatrical situation, nor of the presence of the audience, consequently he/she does not change his/her deictic axis (his/her "I"), a change which, in turn, is the feature of asides addressed to the audience directly. So, in fact it may be claimed that they are soliloquies uttered within a verbal context of a dialogue, without, however, the interlocutors being able to hear the "inner voice" of the speaker. These soliloquies are uttered in a different time stream, which explains why their temporal value does not have to be related to the actual time needed for their utterance. If, in stage practice, the same utterances are addressed directly towards the audience, then they become true asides. However, in the passage used as an example, there is no need whatsoever to address the audience, and the King's speeches remain soliloquies integrated within the flow of a dialogue. Because they are interwoven within a verbal context, they may reveal the features of false asides.

It seems plausible to distinguish between a "dialogic" soliloquy and a "dialogic" aside:[16] the latter is linked to what verbally precedes and what follows it, often comments on the other speaker's words, and frequently allows the speaker to say something audible, though indirect, to the audience, which, for some reason, he or she is reluctant or unable to say aloud before other fictional figures. So, that kind of utterance, which often takes the form of "staccato asides" (as in act 2, scene 3 of *Troilus and Cressida*, where several figures mock in asides everything that Ajax says: there are at least nine asides between the lines 201 and 221), may be treated as a "concealed" part of the ensuing dialogue, its fuller version, and is characterized by brief, rapid comments, often ironic and contradictory to what is being said "officially".[17] For me this is one of the conventions of a dialogic utterance, and not an aside proper. Only when these inserted bits are uttered directly to the audience, with the actor trying to a establish phatic function with the spectators, and no one else can hear them, we may treat them as true asides. When the dialogic asides are not linked to what is being said in the ensuing dialogue, than they may be labelled "detached asides", but again they may be addressed to the audience directly or spoken as if to oneself or to a third figure on the stage. In the former case, they become true asides, in the latter their orientation is towards the soliloquy or dramatic dialogue. Also, in the case of the former, the fictional time of the figure seems to merge with the time of the actor (resulting in an illusion of one present time), whereas in the latter it does not, and flows with a typical gap between the two present times. Let us also note that there are instances of the reverse situation, when a dialogue suddenly springs up within a single soliloquy. For instance, in *The Winter's Tale* Leontes's soliloquy opening act 2, scene 3, is interrupted by the appearance of a Servant in line 9; after brief exchanges of dialogue, the Servant leaves in line 18, when Leontes continues with his interrupted soliloquy. At the end of act 4, scene 4 of *The White Devil*, Flamineo begins his soliloquy (at line 113), which is interrupted by the appearance of Brachiano's Ghost (following line 120), and consequently the soliloquy evolves into a direct address to the Ghost (lines 121–137), and after his exit back to the previous mode (lines 138–146). Similarly, we have dumb shows inserted into lengthy monologues by Gower in *Pericles*,[18] the function of which is to elucidate what is not clear in the verbal account, and also to add a visual aspect to a monotonous speech.

Temporal and spatial aspects

Reason prompts us to say that a soliloquy is different from an ordinary monologue and from an aside, but it is by no means easy to explain in what ways.[19] Let us try to compare the three types of utterances through their employment of time structures. First of all, it seems that the time continuum of the stage action (time) is usually suspended (or altered) in the case of soliloquy, whereas it continues at its normal – within the fictional world – pace in the case of a monologue (time is shared by more than one figure); a monologue is always a part of larger verbal unit, such as a dialogic sequence. In the first instance, however, the suspension does not necessarily mean that time stops: on the contrary, although a soliloquy may appear in a time gap, in timelessness, it may in fact mean the passage of, say, a day, a week, or five hours. It does not need to be defined in time, or "translated" into a measurable unit of time. It usually denotes the state of mind of a figure, its awareness, in a given situation, or provides additional information about events past or those to come, and not an actual utterance in the fictional world (phenomenologically, it is the real actor who speaks, whereas the figure "thinks"), in a specific space and measured time. There is no defined time passed between the beginning and the end of Hamlet's soliloquy "O what a rogue and peasant slave am I!", and it is by no means clarified by the stage action. For this reason, the figure's location in space is also often irrelevant, and remains unspecified, especially in soliloquies which constitute a whole scene, as in *Timon of Athens* 4.1, where Timon utters a soliloquy, forty lines long, where he rails against the Athenians and includes not a single reference to time or location. Similarly, in *Cymbeline* Posthumus presents a lengthy soliloquy constituting a whole scene (2.5), without any indication as to where he is or to the time of the day or night. Thus, the identification of the exact location is not a rule here. Also, the words uttered are not always directly connected with the immediate plot or action (although in its simplest variant they reveal a figure's intentions or plotting), but serve the function of adding depth to the psychology of a figure (in soliloquies they always reveal the truth), which strengthens the motivation and verisimilitude of its further actions and reactions, as, for instance, in *The Two Gentlemen of Verona*, where in 2.4 the Jailer's Daughter gives us an insight into her emotions, complemented by yet another lengthy soliloquy of hers just two scenes further, and, again in 3.2 and 3.4.[20] Even if the location is given (as in *Cymbeline* 4.1, where Cloten begins his soliloquy with "I'm near to the place where they should meet"), the time implied is not very precise. Thus, if in a given production the director has some other figures present on the stage and hear the soliloquy, apart from the consequences of eavesdropping within the development of action, the soliloquy is given a more specific time and space dimension, shared by several figures. This may, of course, be necessary for the needs of particular production, but we must not forget that the primary function of a soliloquy is the increase of the audience (and not the figures'!) awareness of new aspects of the drama and insights into a figure's psyche that are not apparent and do not stem from ensuing dialogues. The constant juxtaposition of the levels of awareness marks the basic strategy of the playwright and the director to create suspense, provide explanation for particular decisions, to create dramatic irony and to add verisimilitude to the development of action (in mimetically oriented plays). This is one of the reasons why Hirsh's theory does not hold ground, because by making all soliloquies audible to any of the figures means that the level of awareness between particular figures and the audience is not differentiated sufficiently to fully control the reception.

Moreover, if alone, the figure presenting a soliloquy may be presented not only in timelessness, but also in unspecified space: in point of fact the space does not have to be physical and three-dimensional, for it is often the psychic space of the mind that is theatrically presented to us. Take two examples of soliloquies: one is Petruchio's utterance given in act 4, scene 1 of *The Taming of the Shrew*, beginning with "Thus have I politicly begun my reign" (188) and ending with " 'tis charity to shew" (211). The speech itself does not contain any references to the exact location, but since this is a continuation of a scene that we have been watching, the stage has not been semantically "zeroed" or reset, and continues its previously assigned meaning. Furthermore, just before Petruchio enters the stage, his servants exchange comments on his whereabouts, and one of them notices their master coming: "[Curtis] Away, away, for he is coming hither" (187). Without that line (which could be cut in performance), we would have a new scene, which could be located anywhere. Even though the physical space is clearly defined, Petruchio presents a proper soliloquy. Now, in act 2, scene 6 of *The Two Gentlemen of Verona*, we have Proteus enter an empty stage, thus beginning a new scene, and presenting a soliloquy (43 lines long), after which he exits, and a new scene begins with the entrance of Julia and Lucetta (in a different and specified space). There is no indication whatsoever as to where exactly Proteus has found himself. The stage in the previous scene involving Speed and Launce has been identified as "a street in Milan", but with the exits of the two figures, it has become semantically neutral again. So, the entrance of Proteus should convey some hints as to where we are. Since – quite contrary to the previous example – it does not, the space remains undefined, and with equal ease could be treated as "somewhere in Milan" or "inside Proteus's mind", wherever he happens to be. So, as we can see, soliloquies can be uttered in both defined and undefined spaces (actual or mental), although the latter has a greater strength to detach the speech from the time flow within a specific physical space, which is "translated" into the mind of the speaker.

The frequent timelessness of the soliloquy may be linked to space: if what we are hearing is the sign of psychic processes rather then actual speech, than the body of the speaker does not have to be visible to the other figures: usually they signal total indifference to the speech which is often passionate, "full of sound and fury". Even if they notice the presence of the speaking figure, they do not notice the act of speaking. When shown on stage in an actual production, this means that the body of the speaking actor is not a sign of the body of the speaking figure; otherwise, others would have noticed him/her speaking. The difference here with an aside rests in the fact that for the duration of an aside addressed to the audience the speaking figure becomes invisible and inaudible to the remaining interlocutors. Crossing the fifth wall by the speaker of an aside means that it is erased from their perceptive powers, which is not necessarily the case of soliloquies, where not the physical presence but the mental processes are the key issue.

There is a technical problem involved when the scene of the kind described above is performed not on an Elizabethan stage, but on an Italian one (i.e., on a proscenium stage), with scenery at the background (a cityscape, for instance): if the director wants the figure to appear in an unspecified space, he/she has the solution of having the actor distance him/herself from the stage set and present the soliloquy from the proscenium or some other location within the interior of the theatre (e.g., from one of the aisles). Lights may also be used for foregrounding the speaker's face and/or body, and leaving the remaining space in the dark. Anyway, we may subdivide soliloquies into those uttered from within a defined space and those which present themselves in timelessness and physical spacelessness. The first of these shows the speaker detached from the world around him, with which he/she is contrasted, the second is a theatrical presentation of processes occurring inside the speaker's mind often in an unspecified location and time. All of this also implies that the actual behaviour, movement and facial expressions of the actor may cease to indicate the behaviour, movement and facial expressions of the fictional figure and become a sign of its mental and emotional turmoil. So, if during his/her presentation of a soliloquy an actor is taking his/her clothes off or is taking a shower or climbing a ladder, and if the time indicated within the beginning and the end of the utterance is four hours, it does not mean that the fictional figure has been taking the shower or climbing the ladder or undressing for four hours: the latter act may be a metaphoric indication of its inner transformation, as if changing one's skin. This also implies that in theatre the actor's body does not have to be a sign of the body of the figure throughout the performance – its sign function may be altered.

Similar are the features of songs presented during the performance and addressed directly to the audience and without anyone else on the stage noticing them. These may be treated as musical soliloquies, for they also reflect a mental state of mind and emotions rather than actual verbal utterances (since nobody within fictional realm can hear them and nobody is conscious that something is being sung). All sorts of variants are possible here, especially in musicals and the opera: in those forms we can have songs mixed with dialogues that are not sung, dialogues composed of sung passages, etc. The convention is often employed in films, especially musical films, where people sing and dance to express their emotions and usually nobody in the fictional word notices that someone is singing and dancing. Characteristic here is the breach of an ordinary flow of dialogue with music, dance and song, followed by a return to the same dialogue as if nothing had happened. Brechtian songs will, of course, tend to break away from the fictional level, and present themselves as musical asides rather than soliloquies. What this means is that in either case the temporal value of the song does not have to reflect the time that passes within the stream of fictional time, and in the case of Brechtian songs, the time of the figure will coalesce with the time of the actor and spectators. It also means that physically the singing figure has disappeared from the fictional world accessible to other figures, which explains why they cannot see it. The case is different when songs are addressed to listeners on the stage, which means that there is no shift in time and space and no suspension of the fictional time. Also, the physicality of the figure is perfectly clear to the others.

Thus, the stage speech as used in soliloquies is not necessarily a sign of linguistic activity beyond the fifth wall. Therefore, the temporal gap between the distinct streams of time, the figures' and the actors', will be often signalled, and will remain consistent within the fictional level. In the case of the aside, the situation is different: the temporal dimensions of the actor and the enacted figure, hitherto distinct, seem to merge. Hence the effect or even the illusion of the actor and figure being one and the same being, speaking to us as if for him/herself (as hinted above, in most cases their deictic axis changes),[21] and thus seemingly sharing the time and space with the spectators. The latter factor (i.e., space) means that while uttering the aside, the actor/figure signals his/her awareness of sharing the space with spectators, that is, of being in a theatre and participating in a performance. However, the signal does not have to imply that the participation is not feigned. Both of the modes of utterance may be explained only through convention.

The length and style of an aside may, of course, differ from one instance to another, but nonetheless it always marks a break in what had been a continuous flow of verbal exchanges. A soliloquy does not need that; on the contrary, a longer break in a dialogue would leave the remaining actors "unemployed", waiting for their cues. The latter situation is of course possible in experimental theatre, or in theatre that does not attempt to follow the rudimentary rules of "real" life, but this would be a lengthy aside rather than a soliloquy. So, in other words, a soliloquy is not a breach of a dialogue or a substitution of one stream of time for another, but remains in its own time, hitherto signalled, and – when its length requires this – is usually uttered in solitude.

As mentioned above, it may, in some cases, take the form of an utterance with "false asides". It may or may not be noticed as "silence" by interlocutors. An example of the former case is found in *Henry VIII* where in 3.2 Wolsey's rather lengthy "asides" are commented on as indicating his internal psychic turmoil by the others:

> [Wolsey's first "aside", ll. 85–90]
> *Norfolk*:   He's discontented.
> *Suffolk*:   May be he hears the king
>   Does whet his anger to him.
> [Wolsey's second "aside", ll. 94–104]
> *Norfolk*:   He's vex'd at something.
>   (91–s104)

It is obvious that what we hear as a verbal utterance remains silence for the other figures, although they do notice that something is going on in Wolsey's mind. This example also shows that there is no breach of time within the fictional stream, experienced by all of the figures present, which is not the case of "true asides". The present time of Wolsey is not the present time of the actor who is playing him (they are, after all, divided by several centuries). This means that the asides are addressed to himself and may therefore be treated as short soliloquies. This also proves beyond doubt that in many cases soliloquies were not signs of an actual speech within the created world.

There are further differences. For instance, in stage practice, a soliloquy does not have to be, and often is not, addressed directly to the audience, whereas an aside has to be directed in that way for it has to stress the reality of a speech act and to establish a phatic function with the spectators. If it is not directed and addressed towards the audience, than it becomes a soliloquy (a sign of a mental process). This means that the same verbal utterance may in stage practice, in two different productions, become either a soliloquy or an aside (as, for instance, Cordelia's single line: "What shall Cordelia speak? Love, and be silent"). The former is a sign of psychic turmoil, the latter a real verbal utterance directed and addressed to the audience (what is still not certain is the reality of the speaker and of the addressee of his/her utterance). Again, it is up to the director to decide which of the options is chosen. In the same way, the director may decide whether a given aside is to be addressed to spectators (true aside), to him-/herself (soliloquy) or to another figure (false aside). For instance, in act 3, scene 2 of *Hamlet*, after the play-within-the-play, the Prince is conversing with Horatio, and later with Rosencrantz, Guilernstern and Polonius: at one point he says, "They fool me to the top of my bent" (l. 375). Now, this could be either addressed to the audience directly, to himself or to Horatio. Similarly, non-verbal asides are often the result of decisions undertaken by people responsible for the actual production and not necessarily by the playwright. These take the form of gestural and facial signals, by which the actor implies that he/she is fully aware of the theatrical situation he/she is in, and that he/she lives in the same time and space as the spectators. So, the effect is roughly the same as in the case of verbal asides, and in both cases a phatic function is established between the actor and the audience. And it is obvious that we cannot have non-verbal soliloquies, although silent monologues are a possibility – it suffices that other figures signal that they hear and react to a monologue that the audience does not hear.

In stage practice, a soliloquy may be presented in the presence of other figures who do not hear it and are engaged in some stage (usually non-verbal) activity, which implies that the mental processes are taking place in the speaker's mind during, say, wedding preparations or over dinner (in recent productions during a soliloquy other figures often freeze).[22] In that case, the time of mental activity of the "speaker" is related to the signalled time of the fictional background action on or offstage. This explains why a three-minute soliloquy may "occupy" almost the entire time of family dinner, if we use the example (which, in turn, allows the author to shorten the stage time for the whole scene involving dinner: it may occupy the time of the soliloquy only and does not require a mimetic dimension). However, owing to the fact that during the soliloquy the fictional present of the flow of action is often indeterminate or suspended, the result might be that the utterance becomes as if "more real", closer to the time of the spectators (i.e., they perceive the time of the utterance and not the time of background action). And, of course, there are numerous examples of soliloquies that are so interwoven into the stage action that their time is congruous with the time flow of what precedes and what follows (as Angelo's soliloquy in *Measure for Measure*, following the message brought by the Servant that Isabella has arrived, and preceding her entrance in 2.4.19–29).

In the soliloquy opening *The Revenger's Tragedy*, Vindice, holding a skull of his beloved lady, long dead, presents all major characters, who in a dream-like procession pass over the stage with torchlight. Apparently they do not notice his presence, nor do they hear his words: we understand that what we see are not real events and people, but – with the exception of Vindice himself and the skull – mental processes occurring in the speaker's mind. This is corroborated by the appearance of Hippolito, who does not notice anyone except for his brother, nor does he see the torchlight, nor hear the words; he does, however, notice the skull. His first utterance confirms this: "Still sighing o'er death's vizard?" (l. 50). Please note that Hippolito uses the verb "sighing", and not "lamenting" or "speaking", which implies that Hippolito sees Vindice while he is delivering his soliloquy, but does not hear it; nor does he see his brother opening his lips in the process of speaking. This leaves no doubt that what we have been watching are not "true" events but mental processes exteriorized (which was and still is a well-known theatre convention).

Who's there?

The question remains, who exactly is it that articulates an aside and who exactly is the addressee? On the surface it seems that it is the real actor (human being) and the real audience. But at least some of the spectators are aware that the actor has these utterances, seemingly improvised and spontaneous, written in his/her script, and that all asides are an integral part of that script and create various relationships with what is being said and done on the stage; furthermore, we are aware that a different audience is addressed every night, so the utterance is not spontaneous in nature. Is it the figure, then, the created fictional "character", who is speaking as its "other self", or, rather, from its different world, making an instantaneous jump in time and space, crossing the fifth wall? Sometimes this seems to be the case, because more frequently than not in asides the *deixis* of the speaker does not seem to change (his/her "I" remains the speaking person's "I" although there may be exceptions to this rule – we should note that in Cordelia's aside, quoted above, she refers to herself in the third person). The only difference is that the deictic axis of the speaking figure seems to be congruous with that of the actor who is impersonating that very same figure because in accordance with the logic of theatre performance it is only the actor–human being who can see the audience and address it directly. So, the effect is as if we had two in one.

This means that either the rule of accessibility has been violated or that the audience (along with the actor) has been incorporated into the ontological sphere of the fictional world. But this leads to yet another discrepancy: in theatre the actor has to signal a deictic axis different from his/her own in order to be able to impersonate someone else, that is, a fictional figure; once the figure's and his own *deixis* merge, the actor seems to be impersonating a figure who seems to be impersonating the actor who is impersonating him or her. This temporal and ontological confusion creates an illusion of a real speech act directed by a real human being to a real audience. So, the phatic function is established between the audience and both the actor and the figure he/she impersonates. A soliloquy does not create that illusion (even if the figure seems to be directing its speech towards the audience, it does not imply a jump in time and space: its *deixis* does not merge with the actor's). The speech is directed towards the audience, and not addressed to it. If, however, the figure notices (or even seems to notice) the spectators, it means that the spatio-temporal border dividing the audience from the stage world is violated. Seen from a different angle, a question may be asked: is it addressing a real audience, in the sense of individuals gathered in the theatre on a particular night? To retain the logic of the theatre art it seems that it is addressing an audience who in fact becomes a sign of an audience rather than a real audience.[23] This would imply that the existing real audience, empirically different on each night, is "read" by the figure as an audience in its own time and space (which is not the time and space of stage action!), empirically fictional, hence the same on each night. This convention enables a fictional figure to address an audience directly. Of course, we must not forget that it is the real actor who is signalling the physically non-existent figure's utterances and behaviour.

Let us notice that in the rare case when the actor–human being stops playing (impersonating) and spontaneously says something to the real audience (provoked by the moment, e.g. quieting down a boisterous audience, or a warning shouted at another actor who is in danger of getting hurt by some accident), the words that he/she uses are not part of the dramatic text and do not belong to any structure existing or foreseen prior to the performance. The audience he/she is addressing are the real people gathered on that evening to see that particular performance (on a subsequent night, with different people and on a different occasion, the "real" and spontaneous aside most probably would not occur). That, however, is not an aside. An interesting case occurs in *The Midsummer Night's Dream*, where (in 5.1) Bottom, playing Pyramus in the play-within-the-play, reacts spontaneously to Theseus's comment on the development of action:

> *Theseus*: The wall, methinks, being sensible, should curse again.
> *Bottom*:   No, in truth sir, he should not. "Deceiving me" is Thisbe's cue.
>   She is to enter now, and I am going to spy her through the wall.
>   You shall see it will fall pat as I told you. Yonder she comes.
>
>   (179–182)

So, within the fictional space and the play-within-the-play, this is a spontaneous breach of fiction and illusion, and Bottom's words are not a proper aside, because they are not part of the scenario and the artistic construct of the whole play (i.e. the play within). Bottom does not speak as a fictional figure pretending to be its actor, but, having crossed the fifth wall, as a "real" human being, as himself as Bottom. This does not create a new fictional time and space, but simply undermines the one that was being clumsily created on the stage. However, from the point of view of the real audience watching Shakespeare's play, this is not an aside at all; neither is it an address *ad spectatores*. On the contrary, it is a dialogic utterance, occurring within the fictional reality of the court of Athens. The seeming aside is created by the fact that the Athenians are watching the same play within-the-play that we, the real spectators, are watching in a real theatre, which creates an illusion of coalescence of time and space, between the spectators on stage (who are actors) and us, in the auditorium. This makes an illusion that Theseus is "one of us", and, consequently, Bottom's words are addressed to us, being the (tonight's) audience.

So, even if we assume that an aside violates the boundary separating the figure from the audience, we notice that this is not done thoroughly or entirely, so to say. An aside breaks away from fictional present time and is seemingly uttered in the time and space of the auditorium, but not thoroughly or entirely. This is a convention by which we are led to understand that what we see and hear is actually an utterance presented by an actor–human being who at that moment is enacting a split or, rather, double role: he/she is playing a fictional stage figure who is an actor playing in a fictional performance, impersonating another fictional figure who (i.e., the latter two) seem to be inseparable (and therefore the "actor" has to be fictional, for a merger of fiction and reality is an impossibility). So, the real actor is playing a fictional actor who is united with the figure he is playing (but only for the duration of the aside). The latter by definition is playing in front of a fictional audience. And it is that fictional actor-figure who addresses a fictional audience. Let us note that the deixis of the real actor remains different from the deixis of the fictional actor (as in the play-within-the-play).[24] So, a different deictic costume is in use. Even though their "I" seems to remain the same, the full merger is impossible because of the temporal and spatial discrepancies: the "now" of the fictional actor is not the "now" of the real actor and their sense of "here" is also different. The real actor is only a sign of the newly established fictional actor-figure (iconically, a perfect mirror-image) and we, the members of the audience on the particular night, become a sign of a fictional audience of a fictional performance. This creates also an illusion of coalescence of time and space, physical and fictional: the actor's and the figure's present times seem to be, but in fact is not, our, the spectators', present within the same stream of time. That explains why an aside may be uttered on each night before a different audience (of course, on a fictional level, the audience remains the same on each consecutive night). This is possible because in theatre different signifiers may denote the same thing, and one signifier may denote several different things. This also explains why the time stream of the plot (dialogue) is suspended, and why the other figures engaged in the scene do not notice anything: they cannot notice an aside, because it is not their interlocutor who is uttering it, but a new figure (a fictional "actor") who "lives" in a different time stream which is allowed its momentary appearance on the stage of a fictional theatre: not only do they not hear anything, but they do not see the "actor" who remains invisible to them, for he/she belongs to a different ontology. They do not hear anything being said, but they also do not see anyone say anything, because the fictional actor is performing in a fictional theatre. If we deleted asides and left the time required for their utterance, we would be left with dialogue full of unjustified pauses (which proves beyond doubt that the time of the dialogue is usually suspended during an aside, and it is only the real actors who wait for their cues, not the figures for the latter do not notice a thing).

The case is slightly different when the figure is invisible to others from the outset of the scene or dialogue, as, for instance, in act 3, scene 3 of *The Tempest*, where Prospero appears "on the top, invisible" (17.1) and makes several "asides" (or, rather, soliloquies?) during the ensuing dialogue between Alonso, Sebastian, Francisco and Gonzalo. The fact that Prospero remains invisible to others has to be signalled by stage action, actors' gestures, gaze and so on. A set of signals has to be introduced, so that the spectators understand that the unity of time and space is preserved here (this is not a dream or memorial recollection), and that only magic has the power to make one of the figures on the stage invisible to the others. The physical laws of our world are violated, and Prospero's "asides" confirm the violation. So, the speech that is inaudible to the other figures onstage is the result not of theatre convention, but of the laws of physics that, within the world of the play, are temporarily suspended. In this particular case, the aside unveils a different convention by which that phenomenon is possible in theatre. This would also imply that Prospero remains himself and is not playing the fictional actor who plays Prospero. This is consistent throughout the play, for even at the very end Prospero appears in his final speech not as a member of the acting company, but as himself, as Prospero who seems to have a life independent of fictionality. Please note that in the same scene, following the appearance and exit of Ariel, Prospero presents a soliloquy, beginning in 83 with "Bravely the figure of this harpy hast thou/ Performed ... ", which ends with his exit at 92, with the rest of the figures remaining on the stage, and yet hearing nothing.

Both an aside and the soliloquy should be differentiated from a specific form of an address *ad spectatores,* such as a prologue or epilogue, for the latter do not share the time and space with the remaining figures inhabiting the fictional realm, are not integrated with the fabulous level of the performance (they are not the result of a cause-and-effect continuum), do not mark a suspension of verbal exchanges or dialogue, do not involve a change in the speaker's *deixis*, marking its return to the previously assumed one, and may usually be treated as the "author's voice", although his/her mouthpiece is usually an actor playing a member of the particular acting company. In the case of prologues and epilogues, there is no marked juxtaposition of the fictional time of the performance and the real time of the audience. The present time of the speaker is the time of the spectator. They seem to be actual speech acts addressed to the audience and not signs of mental processes or speeches addressed to the speaker him-/herself. They do not seem to be a sign of speech acts within the fictional world. Utterances of that kind usually appear as if outside the fictional world, and consequently outside their stream(s) of time. However, both the past and the future times of the speaker and the audience do not coalesce. This shows that in analysing temporal aspects of stage speech, we should take into account not only the present time, but also the past and the future. It may therefore be said that the stage speeches that mark their distance from the time of the audience by their distinct relationship to the past and the future only, are of a special kind reserved for prologues, epilogues, choruses and such like.[25] Thus, speeches made by Gower in *Pericles*, mentioned above, and other prologue/epilogue-figures should not be treated as soliloquies, even though they are delivered in the absence of other figures on the stage. Simply put, they are not contrasted with other modes of speech (and ontology) of the same figure in other scenes of the play.

So, the distinctive feature of an aside would be the concomitant suspension of the stream of time, in which verbal exchanges have hitherto been held. That factor is not a *sine qua non* feature of prologues and epilogues (a chorus is sometimes taking part in an action involving other figures). An interesting example comes, again, from the *Tempest*, where Prospero is presenting the epilogue as himself, and not – as was the convention – as an actor, a member of the company of players (usually asking for applause). Conversely, in *The Midsummer Night's Dream* Puck appears at the end, and addresses the audience directly, although not as a fictional figure, but as the "real" actor who until that moment had been impersonating Puck: now he has ceased doing that and is playing not really himself as a human being, but an actor who – let me repeat – had been impersonating Puck, but is not doing that any longer. This means that the real actor is now playing someone else – it is his second role, that of a fictional actor (who is not playing Puck any longer). Similarly, the actor playing Feste in *Twelfth Night* ends the play with a song, but he implies that he is not Feste any longer but an actor who until then has impersonated the clown ("But that's all one, our play is done,/ And we'll strive to please you every day"). He has ceased to play a fictional clown, and started playing a fictional actor taking part in a fictional performance (so, we have a change of deictic axis here, concomitant to the shift in time and space). As mentioned above, the case is totally different in the *Tempest*. By addressing the audience directly, Prospero makes a rapid jump in time and space, but without the signalled change of his deictic costume, he remains the same fictional figure and seems (by some magic) to be taking part in a theatre performance. However, in both the cases described, we understand that the performance we are witnessing is one in a series, so in fact Prospero and "Puck" are performing not in a real theatre, but in a fictional one, and they are addressing not the real audience, but a fictional one.[26] This is corroborated by the fact, discussed above, that what distinguishes Prospero's time from the time of the real audience is predominantly its relation to the past and to the future.

The fictional level of an aside is additionally proven by the actual wording, which is integrated with the dramatic structure of the performance and with its other, non-verbal elements, and therefore does not fulfil the rules of natural speech. An additional effect of an aside is a violation of the initial agreement between the performers and spectators: we are reminded (as always by conventions) that what we are witnessing is not a scene from real life (it is not a real actor saying something as him/herself), but an artistic structure, an act of communication, a play, which is based on mutual agreement between performers and spectators. Thus the basic difference between a soliloquy and an aside lies in their immediate situational and verbal context and in the addressee of the utterance: all verbal utterances that are not addressed to the audience (but may be directed towards it) and usually are not audible to other figures, are soliloquies and ought to be treated as signs of mental processes rather than actual speech acts (unless the figure is known to have the habit of talking to itself, as Henry in Samuel Beckett's *Embers*, which is a radio play), whereas all verbal utterances which are addressed directly to the audiences and are not audible to other figures, are asides and ought to be treated as quasi-actual – although essentially conventional – speech acts.[27] Soliloquies are to a large degree independent of dialogue, whereas asides are always set within dialogic utterances involving more than one speaker (or constitute a commentary on a situation enacted by other figures). In the first instance, the spatio-temporal boundary separating the two worlds is not violated (the actor and the figure preserve their different deictic axis), whereas in the second case it seems to be violated (but only seems). In the first instance, the actor's body, gestures and movements are not necessarily a mirror-image of the body, gestures and movements of the fictional figure, and his or her face is not necessarily a sign of actual facial expressions of the fictional figure. The bodily and facial expressions may be a sign of mental turmoil rather than anything else. In the instance of an aside, the case is totally different. The actor is impersonating a new type of figure, confusing in its ontological status: for – let me repeat – the real actor is impersonating a fictional actor, performing in fictional theatre before a fictional audience, impersonating the same fictional figure that he has so far been enacting (the confusion is caused by the fact that the real actor's deictic costume seems to merge with the fictional actor's *deixis* – they seem to be one and the same, but in fact are not). Thus, in spite of all similarities, the difference between a soliloquy and an aside may be explained in terms of the different ontologies, times and spaces (and different direction of the utterance and its immediate context, verbal or situational or both) that are in constant play during a performance. The theatre constantly shows us the "stuff" it is "made on", thus revealing its aesthetic function and presenting itself as an artistic composition or construct.

9

Theatre-Within-the-Theatre, or Time-Within-the-Time

Layers of time and space

Of the many definitions of the play-within-the-play, let me quote a very general one, proposed by Manfred Pfister: "In a play-within-the-play one group of dramatic figures from the superordinate sequence performs a play (the subordinate sequences) to another group of figures."[1] On the whole, this definition includes most of the instances of the scenic phenomenon called theatre-within-the-theatre or play-within-the-play, but not all. For instance, in Thomas Kyd's *Spanish Tragedy*, discussed in detail below, the performers (actors) in what at first sight appears to be the play-within-the-play, are not from the same world as the fictional spectators on the stage (who are two in number: a ghost and an allegorical figure). They belong to different ontology and are therefore separated not only by convention, but by the laws of physics, as presented in the play, and it seems that the play shows, in its macro-scale, a situation different from the "classical" theatre-within-the-theatre convention. And yet, in the play, we find scenes, as in act IV, in which the convention is fully realized. In order to define the complexities of the theatrical situations in the play, one has to reconsider and redefine the traditional understanding of the convention. Again, time structures will appear indispensable in that procedure.

So, let me propose a slightly different approach to the theatre-within-the-theatre convention,[2] in which I shall try to prove that it is not only the fact that someone presents something "theatrical" to someone else, but the ability of fictional actors to create fictional temporal structures within, recognized as such by the scenic spectators, that allows us to define the complicated structure of the phenomenon under discussion. This is why the play-within-the-play in *Midsummer Night's Dream* does not really succeed in creating a theatrical act of communication, and in fact shows us the mechanicals' inability to create theatre as art (of course, to the real audience, this inability is perfectly theatrical and is the source of the humour). It should therefore be treated differently from other plays-within-plays, in which rudimentary features of theatre are observed. The mechanicals are funny not because they produce a funny play, but because they cannot stage a tragedy properly (futile effort is often the source of the comic). And the latter is shown by their inability to create fictional time structures, which, from the point of view of the scenic spectators, they constantly transgress, violate and annul.

Similarly, when a play enacted by fictional figures is performed without the physical presence of a fictional audience, it is not possible to talk about a theatre-within-theatre situation. There is no one on the stage who can perceive the performance as theatre, and no one visibly reacts to it. There is no fictional mental retort that is indispensable for theatre communication. Even if the actors on the stage imply through their acting that they are in fact playing before an audience different from the one that is real, without the physical presence of the former, the effect is not the same. What is essential here, therefore, is the factual presence of the fictional audience on the stage, which enables us to observe the ways in which the inner play is comprehended so that we can see how the stage input spaces are blended by the stage spectators. What is needed is a signalled juxtaposition of at least two models of perceiving reality. And the resultant cognitive blends, as signalled by the actors, may be compared with individual blends in real spectators' minds. Moreover, these may become new input spaces, in the ongoing cognitive process of conceptual blending. This is exactly the case in *Hamlet*, where Claudius's interpretation of the play within becomes a proof of his guilt to those of the spectators who have the knowledge and suspicion of the crime (being by itself an input space). Claudius's visible reactions are interpreted as signs of a cognitive process, by which the King betrays the existence of a memorial space, being the scene of (real) murder. A performance of the *Murder of Gonzago* would not make sense without Claudius and Hamlet's being present as spectators, with their reactions clearly visible to the real spectators. Thus it is almost a rule that during the performance within, the figures-spectators, seated in the auditorium on the stage, comment upon what they hear and see (their silence may also be meaningful). Interestingly, their commentary, in order to be audible, has to be uttered in the usual type of stage speech, which on this occasion becomes the sign of whisper, which is yet another theatre convention.

Of course, we, the real spectators, have the ability to perceive the performance, but even if we are treated as the fictional audience, the result is not different: what we experience is that we are watching either a rehearsal or a play performed before a fictional audience, we being the sign of the latter. This is precisely the case with John Osborne's *Entertainer*, where the title role is given to an ageing and no longer funny comedian, Archie Rice. The real actor playing Archie is performing in front of a real audience, but Archie as the figure sees fictional spectators (who are real to him), who are not very enthusiastic about his endeavours; the real audience may, of course, be delighted by the real actor's performance. So the real actor has to achieve something almost impossible: on the one hand, as an actor, he wants to win the sympathy of the real audience, on the other, as the figure, he has to signal the constant disapproval of the fictional audience. Thus, the presence of the audience on the stage is a *sine qua non* of the theatre-within-the-theatre convention, partly because the reactions of that audience constitute a relevant part of the message, often more important than the play within (as is the case in *Hamlet*). It is not enough to have two groups of figures, each belonging to a different time stream: even if one of these streams seems congruous with the time of the real spectators, this does not in itself create a theatre-within-the-theatre situation. What counts is not only what is being performed within, but the reaction of the figure-spectators to it.

Also, the commentary made by the stage spectators has to be co-ordinated with the utterances of the figures in the play within, so that they do not interfere with one another. The commentary enters into various relationships with what is being said and done within the staged play, and thus creates meanings peculiar to the present occasion. It is site-specific and time-specific. In the same way, whatever is said in the play within, may enter into all sorts of relationships with what has until that moment been said and done within the macro-play, and also with whatever will be said and done until the play ends. Thus, the micro-play brings a verbal (and visual) component into the macro-play, which seemingly is not an intrinsic part of the play proper, and yet gains significance during its appearance on the stage and proves that both are inseparable parts of the whole. The peculiarity of this situation lies, among other things, in the fact that the verbal utterances in the play within are not part of any figure-perspectives in the macro-play; however, as it has already been mentioned, they may enter into all sorts of relationships with all the remaining utterances made in the macro-play, including those that are even beyond the perception and cognition of any of the fictional figures (such as songs, soliloquies, prologues, etc.). Thus, the convention may become an important meaning-generating strategy, and not just a scene showing some people enjoying entertainment, being in itself a strong metatheatrical device.

So, in a theatre-within-the-theatre situation, we would expect something similar to Pfister's definition to appear: a sort of multiplied version of a typical theatrical performance. However, it seems that the actual rule is not as simple as that. Indeed, within the fictional realm created on the stage, we would have a theatre performance enacted, which means that in an act of communication, some fictional figures will play the roles of spectators, whereas others will be "double figures" – "double" because they are fictional human beings in the macro play whose profession or hobby is acting, and they also perform in the play within, where they are enacting the roles of figures other than what they are in the reality of the macro-play. Thus, in relation to figures who in the macro-play are spectators, they are actors who are enacting "inner play" figures. This means that time structures must create a temporal hiatus: the stage spectators share the same (biological) time and space with the actors of the play within, they share the same ontology of the same world, but the figures created by these "inner" actors live in a different world, separated by ontology, time and space. Also, in order to create fictional figures, actors in the play within must change their deictic axis: they pretend to be someone else, whose "I" does not correspond to their "I" as "real" inhabitants of the macro-play world.

None of the other figures within the macro-play are "actors", as seen from the perspective within, whereas from the perspective of real spectators of the whole play, all the human beings on the stage are actors, but some of them play double roles: as figures they are also actors in the play within who, in turn, enact roles other than the ones prescribed to the same human beings in the macro world. So, fictional figures from one sphere enact on the stage within figures different from themselves. In other words, the real actor on the stage plays a fictional actor inhabiting the fictional realm, separated from the real audience by time and space, and within that reality he/she performs in a theatre performance, enacting another fictional figure, inhabiting another fictional realm, separated from the fictional stage audience by time and space. It is important to note that the first actor is not the same as the second one. They belong to different time streams and a different ontology. In other words, in the theatre-within-the-theatre situation we are dealing with at least three present times: the temporal present of the real actor, the present time of the fictional actor/spectator and the present of the fictional figure within. The tri-partite present belongs to three separate streams of time. In theory, the inner play may be even set within a time stream reminiscent of ours, that is, the real spectators'. This also means that during the performance within, the differences between the time structures employed will have to be signalled constantly, along with the space markers, so that the spectator without has no doubt as to what belongs to the inner play world (space) and what does not.[3] Please note that the situation described here is different from the one involving asides, as described in the previous chapter, because in the latter case the fictionality is veiled behind the convention; here it is perceived as fictionality by the stage spectators.

How, then, do my earlier observations about actors and acting apply to the play-within-a-play, and the actors within? As stated above, the acting relates to two fictional strata and whom we see is (a) a real human being whose profession or hobby is acting (i.e. the real actor), (b) the fictional figure he/she is enacting, being in this particular case an actor, (c) the latter becoming actor-in-action in the inner play, and (d) the "inner figure" that the "fictional" actor enacts in the play within. And we also see (e) the fictional audience and the stage spectators' reactions to the inner play, which is a part of the scenic communiqué (even if they do not react spontaneously, the fact of their silence or indifference is significant). Consequently, the acting within reveals a dual orientation: the actors perform before two distinct and separate audiences, the one on the stage and the real one. However, the former is also acting and therefore cannot be the addressee of the duality, which they simply do not notice. Besides, the audience on the stage can never be the true audience for also their future is preordained and given.

We also observe all sorts of verbal and visual, syntagmatic and paradigmatic, relationships that the text of the play within creates with the macro-play. Now, the spectators within the play see that differently, for they do not see the show as a play-within-the-play, but as a "normal" theatre performance. For that reason, they see (a) what is a real human being to them, whose profession or hobby is acting, and (b) the fictional figure that he/she is enacting (which reflects the way we perceive them, the spectators of the play within, but not the actors in the play within), which is the same as the (c) or the "inner figure" in the list above. And they do not see any stage spectators watching the play within. So, at least two elements are missing ("d" and "e" from the list). Also, they are not in position to notice the whole network of relationships between the play within and the macro-play.[4] Consequently, the conceptual blending process is different in the two audiences engaged. The spectators on the stage do, however, notice the fact that the figures in the play within do not notice their presence, that they pretend to be in a different place at a different time, and that the time in the fictional world flows differently from what their clocks show. The juxtaposition of the two perspectives, the actors' and the spectators', is significant. Equally important in the employment of the-play-within-the-play convention is the juxtaposition of the reactions of two separate audiences, the one on stage and the one in the auditorium, which is likely to be different for the obvious reason that the levels of awareness of both will not be congruous. This is why the play-within-the-play without the stage audience does not make much sense. Simply put, the convention is composed of at least two elements, which means the show within plus the audience and their reaction to the show. The relationship between the two generates meaning, which cannot otherwise be transmitted to the real audience.

Let me recall here a special case of the staging of the theatre scene in Robert Sturua's *Hamlet* (Rustaveli Theatre, Georgia, 2006). In it, while the actors in the play within are performing the poisoning scene, the King and Queen from the macro-play leave their seats and take an active part in the scene, by mirroring the movements and utterances of the players. This is arranged in such a way that we have two groups of actors, enacting exactly the same movements, blocking, gestures and words, in two groups along the vertical axis of the stage. In the world of *Hamlet* nobody notices the awkwardness of the situation, for the simple reason that the part of the scene played by Claudius and Gertrude is not part of that world, and may be perceived by the real audience only. This means that the emptied chairs of the King and Queen, within the created world remain in fact occupied throughout the show within, and no one has noticed anything bizarre happening. This implies further that what we see is a visual presentation of the "meaning" of the play within, as a recreation of events past. The staging confirms Hamlet's suspicion, and shows us that the staging by the players is an exact enactment of the events as they really occurred. This also means that the part performed by Claudius and Gertrude takes place at a time different from the play within, it is much earlier, and at a different space (old Hamlet was poisoned in a garden and not in a court hall where the performance is taking place). Thus, the denoted meaning of the play within is visualized on the stage as a faithful historical reconstruction or an exteriorization of Hamlet's imagination. The temporal structures employed here are, of course, quite complex, because we have a simultaneous presentation of a seemingly historical event, or a mental process, and the play showing it. However, since substantially and stylistically the historical bit remains theatrical, we understand that this is still part of the macro-play, an exteriorization of Hamlet's, or the King and Queen's psyche, perhaps, and not a real documentary piece of evidence. Thus, the scene may be treated as an original staging of the stage audience's reactions, in this particular case of the Prince's or the royal couple's inner turmoil. The latter might also imply that the Queen was in fact engaged in the murder.

It may also be said that the sign nature of the actors changes, again, owing to the fact that they play dual roles, those of the actors within and of the figures enacted by them. So, as an icon the real actor signifies the fictional actor, in and out of employment on the stage, and consequently the real actor signifies a human being who is real within the fictional reality. During the performance, the fictional actor, now employed, becomes an icon of a stage figure within, and so becomes a dual icon for us, being external spectators. As an index his/her appearance becomes even more complicated, for during the show within he/she presents him/herself as (a) an actor enacting feelings, thoughts and emotions of a fictional actor in the play without (though the emotions of the real actor are usually concealed), who is presented at work and thus his/her behaviour and utterances are also an index of his histrionic skills and talents; this fictional actor also presents him/herself as (b), that is, the fictional actor who enacts feelings, thoughts and emotions of a fictional figure in the play within. So, when the fictional actor cries for Hecuba, his tears denote both (a) and (b), the act of playing as such and an enactment of yet another fictional figure. And that is the distinctive feature of acting within. In particular productions (and in films[5]) the stress may fall on either (a) or (b). The actors in *The Murder of Gonzago* may be shown in such a way that their fear of the consequences of the intrigue against the King will become apparent. In *The False Servant*, being one of the plays within *The Roman Actor*, Paris is aware that the Emperor is enacting a revenge scheme on the stage. In that case the indexical non-verbal signs (such as gaze, facial expression or gestures) will refer to the primary fictional level rather than to the world created on the stage within. Also, in the style of acting in which the real actor distances him/herself from the figure enacted we might have a similar effect. And usually the acting style (along with costumes, make-up, wigs, etc.) in both fictional spheres is conspicuously differentiated. As we can see, this convention provides all sorts of creative opportunities to the directors and actors.

Let us also observe that other relationships become more complex than in the usual theatrical situation. In the case of the play-within-the-play situation, the rules of the theatre applied within are contrasted not immediately with the rules of the non-structured world of the real auditorium, but with those belonging to the structured world of the stage and its fictional audience. This means that the inner performance is given the function of an index sign and creates all sorts of relationships with the structure of the macro-play (e.g. through contrasts in costuming, acting, directorship, time-flow, music, parallelisms in plot development, etc.), and for that reason cannot be separated from the latter and interpreted autonomously. The same rule applies to acting, which in most part relies on its indexical function: actors/figures in the play within do not create relationships with other inner play figures only, but enter into various relationships with the figures in the macro-play (which is not the case in the usual theatrical situation). The King in the *Mousetrap* (in *Hamlet*) is not only a figure in the play within, an index sign that may be related to other figures within, but is also a Player in the macro-play, and creates all sorts of indexical relationships with other figures in the macro-play, both as the King and as the Player (please note that in each case the network of relationships will be different!).

It is worthwhile to comment briefly on the language used in theatre-within-the-theatre. As everything else, the stage speech, discussed in detail above, becomes more complex than in productions that do not employ the convention. Every word enters into relationships with other words, bodies and objects within two strata of fictionality. Syntagmatic and paradigmatic relationships and equivalences may be created involving signifiers that do not share time or space in the fictional realm. Some of these may be noticed by the scenic spectators, while others are not. The costumes, the set and blocking of the actors in the play within may, for instance, allude to a painting known in the outside world, or to a piece of sculpture, or a scene from a contemporary film, or a commercial. It may also allude to earlier scenes in the macro-play. This enables the director to build many-layered semantic clusters, in which every word might have several planes of reference.

Screens, monitors and asides

In my discussion of asides, I have indicated that during the actual utterance of an aside addressed directly to the audience, the fictional figure seems to merge with the real actor. This is why we have an effect of the real actor speaking to us as if he/she were a figure who miraculously jumped over the temporal and spatial hiatus separating the world of the stage from our own. If the relationship between the play-within-a-play were the same as in the "usual" stream of action on the stage, then the above rule would apply. However, it seems that the situation is different, because it is not a simple "doubling" of a theatrical situation: temporal and deictic structures and signals are different in the situation of theatre-within-the-theatre, and the application of an aside might be a good test for the difference. As explained in the previous chapter, in an ordinary theatre situation, during an aside the real actor is playing a fictional actor who seems united with the figure he is playing (but only for the duration of the aside). The latter by definition is played in front of a fictional audience (during an aside, the real audience becomes the sign of a fictional audience). And it is that fictional actor-figure who addresses a fictional audience.

Naturally, the situation is quite different when an aside is made from the screen or monitor, in the case when other media are employed on the stage, as is often the case in (post)modern theatre. Any recording is by definition set in the time past, so an aside made from the screen cannot rely on the same theatrical convention as the one described above: the speaking "I" cannot address the spectators directly and for that reason is not violating the rule of accessibility and the fourth wall. We can only imagine a situation when the speaking agent is presented live on the monitor and is capable of perceiving the reactions of the audience: in this way, even though the spatial distance is signalled and preserved, we are dealing with temporal congruity – the real time of the audience is the time of the performer. In this case an aside works only if the common present time is contrasted with the theatrical dual present; if it is not, an aside becomes an address, which is not a convention. So, the screened actor, or the one in the monitor, has to signal his/her present time, congruous with that of the audience; moreover, he/she has to signal his/her ability to perceive the reactions of both the stage figures and the spectators; also, the situation requires that the screen or monitor are clearly visible to the stage figures, so that an aside may ostentatiously violate the rules of acoustics. All of these conditions are rather hard to meet, which explains why the convention simply does not work in that way.

Now, let us consider a situation, plausible in today's theatre, when the play within is presented as a "live" relay to the spectators on stage. First of all it would be important to notice if the stage spectators perceive what they see and hear as presented on the monitors, or not. Anyway, the situation becomes quite complex as far as the mixed mediums are concerned. Leaving aside the significant feature of one dimension missing, the very fact that someone appears on the screen or a monitor implies a spatial distance, which, in turn implies the lack of spatial contiguity, which severely undermines the indexical relationship between the two sides. This makes any direct interaction hardly possible, even though the rule of two present times might be preserved. How can one make an aside, from a screen to the stage audience, or to the real audience? The difference is that the fictional present is not signalled simultaneously with the real present within one space, which further foregrounds the distance between the two realms. The fictional present time does not overlap the real one, so the duality of time streams is not a blend. The two present times are separated by the two mediums employed. The juxtaposition of the mediums may, of course, be artistically fruitful, but we are not dealing here with the theatre-within-the-theatre situation. This is not a theatre audience, sitting on the stage, but a television or film audience. In a number of recent productions, we have yet another convention introduced, that of a live actor conversing with a screened two-dimensional actor (live or a recording). Again, although the fictional present time is signalled by both interlocutors, they are objectively separated by space, which means that the present time is not created within one space and is actually dual or double in nature – each operates within a different space, which, moreover, do not overlap. And one more thing: since the screened or monitored actor is not presented "live", he/she cannot function as a biological clock, showing the continuum of the real present time, at least not within the space of the stage.

Again, we must ask the question whether the screen or monitor exist in the fictional world on the stage? If they do not exist, then where exactly is the interlocutor? Is he/she in the same space as the live actor on the stage? What if both interlocutors (or more than two) are screened or monitored? In that case, the ontological, temporal and spatial complexity is even deepened, something that the postmodern directors tend to forget. At any rate, the described possibilities make the whole situation different from a traditional, theatrical one, where the fictional present continually overlaps with the real one, generating the effect of theatricality, which covers the whole space of the stage (unless we have two or more fictional times operating). The fact that it is one space is important also for a reason that has not hitherto been mentioned: the spatial contiguity makes the communication between the stage figures plausible – it retains the rudimentary features of a lifelike verbal exchange. When the spaces of the speakers are separated all sorts of questions arise, which undermine the rudiments of theatre. For instance, we may ask whom exactly is the screened person seeing and talking to: a live actor, whom we see on the stage, or his/her relay on a screen or monitor in another space? The signalled lack of spatial contiguity makes the first option improbable, unless we accept the convention that in the fictional world there are no screens or monitors. Thus, the appearance of the latter on the stage might gain metaphoric meanings, such as, for instance, signalling the distance between one speaker and the other, or an attitude of one to another. As I said, audio-visual devices bring all sorts of creative opportunities to theatre, but they cannot dominate, for their spatial and temporal relationships are different and contradictory to those that are essential to theatre.

Now, returning to the main line of thought, let us consider a special and rare case of an aside made in the play within, but not to the fictional audience sitting on the stage and watching the play within, but to the real one, as if the actor in the play within noticed our presence. This would make him/her unique, because usually the stage spectators do not notice anything outside their world. The question remains whether they would notice the fact that the actor/figure in the play within was saying something to a world inaccessible to their senses, that is saying it in some other direction than where they are seated, and not belonging to the fictional realm created on the stage? If they did notice anything, then the aside would be breaking through two fictional layers, the one of the play within and the other of the macro-play. Both are separated from the spectators' sphere by an invisible wall of conventions (the rule of accessibility). Time structures would become very complicated in the event of such a violation of rules.

This simply cannot work in the same way as a conventional aside. For this reason, the actor can only make an aside to the fictional audience on the stage, because their time, space and ontology are the same. A fictional being does not create an illusion of itself, for – quite contrary to the real actor – it is a fictional construct and nothing else. So, the only way in which this could be carried out and explained would be to assume that the real actor is making an aside as if he/she were playing the role of a fictional actor who in the play within is enacting yet another fictional figure. But, as noticed above, one is not the other, and in asides we have to have an illusion of sameness, of the oneness of the fictional figure and the real actor. In the theatre-within-the-theatre situation we would not know which of the two fictional creations were speaking, the figure within or the fictional actor within, and which should be identified with the real actor. It may also be noted that the real audience's awareness of the presence of the real actor in the show within is somewhat weakened by the fact that he/she is "two fictional strata away from him-/herself".[6] Simply put, the convention does not work in the same way as it does in the ordinary sequence of a play on stage. If employed, it would create confusion in spectators, which, of course, may be the goal of the play-wright or director (as in Tom Stoppard's *The Real Inspector Hound*).[7] However, asides made from the plays within towards the fictional audience on the stage work perfectly well, although their function is different from the ones made directly towards the real audience (we observe an aside in action, and do not experience it as a direct address to ourselves). So, once again, time structures help us to explain the ways in which theatre works.

To sum up, the following list of optional audience reactions may be proposed: (1) an aside is made from the inner play to the fictional audience who reacts to it appropriately, which parallels the "normal" use of the convention; (2) an aside is made to the fictional audience who does not notice it; (3) an aside is made to the real audience and not to the audience on the stage who does not notice anyone making an aside; (3a) in this variant the audience on the stage notices the fact of an aside being made, but does not notice the addressee; (3b) in this variant the audience on the stage notices both the aside being made and the addressee (3c) in this rare but possible variant, the audience on the stage notices the real audience's reaction to an aside made from the play within, but does not notice anyone making the aside; (4) an aside is made to members of both audiences and the fact is noticed by the stage audience; (4a) an aside is made to members of both audiences and the fact is not noticed by the stage audience; and optional (5) an aside is made to the real audience by a member of the stage audience (with or without anyone else in either of the fictional strata noticing the fact). We may also imagine asides being made from the screen or a monitor set on the stage. The juxtaposition of the two audiences' reactions may be significant, and may also be the source of humour or dramatic irony; it will also reveal its strong metatheatrical character.

Time, space and magic

Usually, in its length, the macro-play dominates and the theatre-within-the-theatre is a relatively short scene. Owing to the power of theatre to model fictional time, a five-minute show within may stand for a full-length play. That of course undermines the verisimilitude of the scene, but may be accepted with the help of convention. We understand that a short play within operates in a condensed time stream, and becomes a sign of events occurring in a different current of the macro-play. Some playwrights, however, seem to have reservations as to the power of the convention to convey meaning, so they find excuses for making the play within short, such as a sudden interruption of the play (as in *Hamlet* or in *The Tempest*). Philip Massinger, for instance, in his *Roman Actor*, which contains an amazing number and variety of plays-within-the-play, provided several excuses for the "condensed" plays within. Before the play on avarice begins in act 2, scene 1, Caesar says:

> Let them spare the prologue,
> And all the ceremonies proper to ourself,
> And come to the last act ...
>
> (274–6)[8]

In act three scene two, Domitia gives another explanation for abridging the next play within:

> I have been instructing
> The players how to act, and to cut off
> All the tedious impertinency, have contracted
> The tragedy into one continu'd scene.
>
> (131–4)

And in the crucial second scene of act 4, where Caesar is preparing to enter the stage action as a player, he orders the actors to cut the play: "We'll have but one short scene" (233). In the case of *The Roman Actor* the device of explaining the cuts is fully justified, for all the plays within have immediate relevance to the plot and figures in the play without, so the time structures in the two streams ought to be similar if not the same. This also enables the figures from the macro-play to enter into the world of the play within, as in the scene when Caesar becomes an actor and uses his assumed role to implement revenge on Paris (who has become the Empress' lover), who is also playing in the same performance (*The False Servant*). At one point, the Emperor takes his cue, but forgets the lines, or pretends having forgotten them: "O villain! Thankless villain! – I should talk now;/ But I have forgot my part. But I can do,/ Thus, thus, and thus. *Kills* PARIS" (4.2. 280–3); to which Paris exclaims: "Oh! I am slain in earnest" and dies soon after. Let us note that both utterances are spontaneous in nature, and are uttered clearly by the men from the macro-play, and not by the figures they enact, at least from the point of view of the real specta-tors. However, for the spectators in the outer world, the utterances are of the figures, the Emperor and Paris respectively. This example shows the duality of everything on the stage, i.e. the effect of theatricality in action. The real actors simultaneously impersonate both figures who are actors and figures who are not actors. Each sphere is located within its own stream of time, each is sepa-rated by the fifth wall from the phenomenological world.

Yet another device is the dream inset, which often takes the form of a dumb show within the macro-play, but sometimes is equipped with verbal utter-ances of the apparitions. The former may be exemplified by "The Vision" in 4.2 of *Henry VIII*, where Katherine falls asleep (in the presence of some other figures), sees a mimed dream vision, wakes up and finds out that nobody else has seen a thing; an example of the latter may be found in 5.1 of *Richard III*, where Richard falls asleep in his tent and sees not one, but a series of ghosts of the people he has murdered. All the ghosts address the King and, inter-estingly, also address Richmond sleeping in the tent on the opposite side of the stage. This implies either that the ghosts appearing in the mind of the dreamer have the ability to cross over the mental reality and enter the real world, or that the ghost is dreamt by both figures at the same time or that what we are seeing is not a dream but rather a real apparition which has the ability to communicate with the two sleeping men at the same time. Usually a dream inset is the mental process exteriorized, which means it takes place in an amorphic mental space, and in timelessness (unless time is specified), but in the case of Richard and Richmond, if we accept the last option, we do not see an exteriorization of anyone's dream, but a "true ghost" who communicates with both men in their sleep. An even more elaborate dream inset appears in *Cymbeline* (5.4), where not only do the ghosts materialize on the stage and "perform" around the sleeping Posthumus, but even Jupiter himself descends "in thunder and lightning". Similarly, in Massinger's *The Roman Actor*, we have a scene in which Caesar falls asleep and "*A dreadful music sounding, enter* JUNIUS RUSTICUS *and* PALPHURIUS SURA, *with bloody swords; they wave them over his head.* CAESAR *in his sleep troubled, seems to pray to the image; they scornfully take it away*" (5.1. following line 180).

One of the magical features of theatre art, discussed in Chapter 2, is its ability to assign meanings to almost any substance, even to thin air (as in a clown bouncing off an invisible wall). The inadequacy of the substance of stage signifiers signals to us the second model of perceiving reality, that of the stage figures, for whom the substance is adequate. This also means that in the theatre-within-the-theatre situation we may sometimes have more than two models of perceiving reality juxtaposed. The stage spectators will perceive the inner play conspicuously differently from the way the same world and its material representation is perceived by the figures within. And either or both may be congruous or not with the way we, the real spectators, perceive the same phenomenological world visible and audible on the stage. For instance, in the scene mentioned above from *The Tempest*, where a masque is being staged by actors who are spirits evoked by the magical powers of Prospero, seemingly the same actors are both real (from our perspective) and insubstantial (from the perspective of Prospero's world). From our perspective we see real actors – human beings who enact actors who are spirits; and those actors enact mythological deities. The same many-layered meanings may be applied to any objects used in the play within. This enables the playwright and the director to create all sorts of relationships between figures and objects belonging to various strata of the performance.

A special case of theatre-within-theatre convention is to be found in *The White Devil* by John Webster. There are two dumb shows presented in act 2, scene 2, both of which are unusual in many ways, for they are presented to Brachiano by Conjurer as a show of magic, by which events taking place simultaneously to the presentation, but somewhere else, are shown in the here and now of the spectators. They also take the form of a theatrical mime-show, and include elements of scenery (a curtain with Brachiano's picture behind, a vaulting horse), costume (Isabella's nightgown, "spectacles of glass"), props ("perfumes", "lights"), metonymy, and a changed style of acting, specific for dumb-shows ("Enter suspiciously, JULIO and CHRISTOPHERO", "MARCELLO comes in, laments" etc.). However, this is not a summary of events to occur or those that have already occurred, for we are informed that they are taking placing "now", that is, in the time of the presentation, but somewhere else. And the dumb shows are not perceived as a theatrical performance by the two stage spectators. But first of all, the figures in the dumb shows, although three-dimensional, are not presented live, but are a sort of projection from a distant space, presented in the *hic et nunc* of Brachiano and Conjurer. In other words, what they see does not follow our initial definition of the theatre (what we, the real spectators, see, is something different altogether). Instead, what we have here is a form of theatrical "transmission" and by the power of magic we are shown real events from a different space. That explains why figures from the dumb show cannot see Brachiano and Conjurer, although the latter can see them; the theatrical rule of accessibility is explained by the employment of magic. It is the magical "night-cap" that enables Brachiano to take part in the show of magic. It is not, however, a truly theatrical convention, for the figures in the dumb show cannot see their spectators not because of the convention employed, but owing to the power of magic. Thus they are not presented as transgressing the fifth wall, for it is not the latter that separates the two sides. Thus, the alterations and distortions in the appearance ("picture"), which is metonymic and silent,[9] does not seem to derive from theatre conventions, but from characteristics of the medium, which is magic: it stresses the fact that the reality presented is being transmitted, and not enacted as in theatre.[10] The people there are not actors, and are not perceived as such; they do not act, do not imperson- ate anyone but themselves, and are not engaged in any activity that is not part of their own lives. But most importantly, as I have already observed, they are not live presentations (a similar effect today would be if the dumb shows were presented on the screen). Time is not condensed or stretched; it flows with the same tempo as the time of the stage "spectators". This means that from the point of view of stage spectators, we are not dealing here with theatre, but with a show of magic, by which many centuries before televi- sion a transmission of events occurring elsewhere is made possible. From the point of view of real spectators/readers, however, the situation is different: the presented events are real only in the sense of being part of the perform- ance, and do not refer to anything we might consider "real", but are part of stream of fiction that flows on a theatre stage. They are a stylistic theatrical device, by which events occurring off the stage are presented in an unusual transmission, a relay. So, from our perspective it is not a transmission, but a sign of a transmission within the performance. Webster could have staged the same events in two successive scenes, but he would lose two advantages of the dumb-show: first, he would lose time (dialogues would have to be added), and secondly, he would lose the powerful scenic device, a display of magic in action, which could and probably did amaze the audience.

The missing act

The last example to be discussed in this chapter is a very special one, not only because of its historic significance, but also owing to the variety and complexity of theatrical devices it employs. Traditionally, it has been accepted that perhaps the first major dramatic piece written in English and employing the play-within-the-play convention, in which most of the text belongs to the inner sphere is Thomas Kyd's *Spanish Tragedy* (1589?). In it, the Ghost of Andrea and the allegorical figure of Revenge sit on what seems to be a theatre stage and watch what seems to be the play of "Spanish trag- edy" evolving in front of their eyes.[11] The title of the play, of course, suggests a location, but the appearance of the Ghost and Revenge, in addition to Andrea's expository account of the underworld, bring about some uncer- tainty.[12] In theatre, we like to know where we are, and the division of space into two spheres, the performing area and the auditorium, lies at the roots of the art.

Quite contrary to our expectations, Revenge neither identifies the location as Spain, nor as any other specific place ("here"). He merely says "thou art arriv'd/ Where thou shalt see",[13] and identifies the type of the spectacle to be presented as "tragedy". Where does one sit down and watch a tragedy? The answer seems obvious – in the theatre, and at first sight there is no reason whatsoever to treat *The Spanish Tragedy* as taking place anywhere else but in a theatre. This appears, then, like a classical example of a theatre-within-the-theatre situation. When staged four hundred years ago, or today, *The Spanish Tragedy* has to be presented in a theatre (whatever space that functions as a theatre), and all the actors on the stage, in both spheres, the outer and the inner, signal that the figures impersonated are not in the same spatial and temporal relations to the real audience. What we witness on the stage is that one group of actors performs a play, or what seems to be a play, before another group of actors, and we, being the real spectators, are facing a double-layered spectacle. In accordance with theatre conventions, the figures in the play within do not notice the presence of the spectators without. The latter, in turn, do not notice the real spectators' presence.

However, when we talk about theatricality, or theatre situation (act of communication), we mean that the phenomenon under discussion is based on a convention, on mutual agreement, by which one side, being the actors, agrees to behave strangely and to pretend to be somewhere else and at a different time in front of another group, being the spectators, who agree for a certain period to suspend their disbelief and "read" the performance as a sort of unusual message that someone wants to convey us. The mutual agreement guarantees a successful act of communication. In Kyd's play, however, no one makes any agreement, and even though we have two people watching another group of people who remain unaware of being watched, this seems to be occurring without the aid of theatre conventions, and reminds one of a situation rather, in which by some magic the two "spectators" on the "stage" are given an insight into the lives of real people, well digested into scenes, to use Hamlet's words. Even though the time of these scenes flows with a different pace from the time of the "spectators", we must not forget that both the Ghost and Revenge come from the Underworld, and represent the metaphysical sphere, governed by its own laws of physics. Anyway, since the element of mutual agreement between the two sides involved, the actors and the spectators, is missing in Kyd's play, and no one makes any hints as to a different time and space of the seeming "spectators" and "actors" (on the contrary: the Ghost and Revenge always talk about the seeming actors as real people), one is tempted to question the initial observation, taken by many for granted, that what we are witnessing there is in fact a play-within-a-play.

During the actual performance what we, being the real spectators, see, is undoubtedly a theatrical situation, with actors on the stage, with visible elements of scenery, theatre architecture, costumes, wigs, lights, traps, music and the like. However, it is not that clear that for the figures involved the same space is theatrical at all. The crucial question remains whether the "actors" in the play within (the "Spanish Tragedy") are shown at work, so to say, and enact assumed roles of fictional figures, or whether in the fictional realm they are not players by profession, but real human beings and are shown – by magic – in real-life activities, evolving in front of the two spectators and belonging to the same stream of time (at least within individual scenes). The question is crucial, for it determines the nature of the entire play: is it an early example of a theatre-within-the-theatre convention, by which theatre becomes a metaphor for human life, or is it something else, where real life becomes a metaphor for theatre? If the first option is selected, then the figures we see in the play within are not alive, are only the signifiers of fictional creations, and the play can only be an enactment of something that has already passed or something that will happen, belonging to a different stream of time, which has a marked beginning and an end. The latter fulfils our rudimentary definition of theatre and means that what we see is a group of actors who pretend that what belongs to a time different from the time of spectators, is their merged present time. The split present time is the fundamental feature of theatre art. The second option, however, gives us a different explanation of what is occurring on the stage. By some magical powers, Revenge has the ability to visualize or present the concomitant events occurring in a different space, being a sort of early example of a reality show.

The discrepancy is caused by the fact that the people we see on the one hand seem to be live actors enacting events from the past, but on the other the implication is that they are enacting themselves, and not as actors but as Spanish royalty and courtiers. The Ghost recognizes all the figures in the play within as real, and he never speaks of them as actors; he does not seem to be aware of the theatrical situation he has found himself in. He does not notice any difference between what he observes in the course of action, and what he remembers from his own life led between the very same people and within the very same spaces. Moreover, he treats everything he sees and hears within that space as really taking place in his here and now, and not as an enactment of something already past. He takes all the "actors" as real people. He recognizes the world presented to him on the stage as real, and does not suspect it being a sign of anything else. This implies that from this point of view, the deictic axis of actors and figures remains the same: there is no implied and signalled discrepancy between the actor and the figures who on top of everything else, as already indicated, are not recognized as actors by either the Ghost or Revenge.Their speech is not recognized as verse, their actions and gestures as theatrical. This means that the language used by the figures is not the language that the Ghost and Revenge hear: paradoxically, it is the same language (blank verse) that they use, at least from the real spectator's perspective. This would imply further that the Ghost and Revenge are watching events "live", brought to the stage with the help of Revenge's supernatural powers.

Moreover, the ontology of the actors in both spheres remains different: the Ghost and Revenge are metaphysical, and they remain invisible to the "actors" in the play within throughout the performance. What is important here is that the basic rule and convention of theatre does not arise here. The "actors" in the play within do not pretend that they do not see the spectators of their "play", for the simple reason being that they objectively cannot see them anyhow. They do not have to pretend anything. The laws of physics are different in the sphere of life, and in the sphere of the underworld, where it is possible to cover great distances in a blink of an eye. So, there is no mutual agreement between the two sides involved, and there is no convention employed. In other words, there is no play staged before the Ghost and Revenge, and they are not true spectators in the real theatre. What they observe are real-life activities of real human beings ("real" from the point of view of the figures involved), who do not enact anyone else but themselves. They do not even "act", for they simply remain themselves. From this perspective, what we see is a slice of life, which is presented to us as a kind of reality show in theatrical form, which is not accessible to any of the figures on the stage belonging to the Spanish sphere. In this way, Kyd is creating a metaphor of life as theatre (and not of theatre as life).

The problem is even deepened by the time structures involved. For we are not presented with events from a different current of time, either past or future, which means that if the "actors" in the play within are in fact enacting themselves, they are not separated by time from the spectators without. The distance is predominantly spatial, as in the example of *The White Devil*, discussed above. It is worth noting that throughout the play all references to the past and to the future take the same point of reference, common to the "actors" and the "spectators". This is quite contradictory to the basic requirement of fiction: it is the temporal discrepancy that makes fictional figures out of actors enacting themselves. It may also be observed that within the whole play we do not have a single utterance of the figures in the play within which could be treated as an aside directed towards the fictional audience.[14] In order to have an effective aside, the actors have to be alive and/or belong to the same ontology as the spectators. In the case of the play within, the time of the performers (fictional actors) has to be the same as the time of the fictional spectators in the macro-play; it has to flow within the same stream. Apparently this is not the case with *The Spanish Tragedy*, because it is also the time of the figures within that seems to be, at least within individual scenes, congruous with the time of the Ghost and Revenge; although "life" is presented as if cinematically edited, for it is cut into "slices" being scenes and acts, between which time evaporates. So, time structures alone show us that what Andrea and Revenge are watching is, from their perspective, anything but theatre. What we have here is a juxtaposition of two streams or even dimensions of time, earthly and metaphysical, both of which are presented as factual. Naturally, it remains theatre from our perspective, for the simple reason that the temporal relations are different. Neither the time of the play within, nor the time of the play without, are congruous with the time of actual spectators.

In the course of the play, we can clearly see that Revenge functions as the director-magician of the whole work, and knows perfectly well what is to happen, and has the power to stop or suspend the action if need be. At one point, in order to elucidate the development of the action and to explain the significance of the events to come, Revenge employs yet another theatrical device: a dumb show.[15]

> Behold, Andrea, for an instance how
> Revenge hath slept, and then imagine thou
> What 'tis to be subject to destiny.
> Enter a Dumb Show.

> (3.15.26–28)

In point of fact, various forms of a spectacle is constantly used in Kyd's play as an illustration, commentary on and elucidation of the reality presented. Also, it is worth observing that from the perspective of real spectators what we have here is now a play-within-a-play: we are dealing with three levels of fiction, the macro-play with two figures in it, the Spanish court with the whole cast, and now the dumb show. The latter is a mime show, but certainly has the ability to mark the temporal and spatial hiatus, separating it from the play without. Of course, the figures at the Spanish court do not notice a thing: the dumb show remains invisible to them. Thus what we have here is a simultaneous employment of two conventions: a theatrical "reality show" and, in the sphere without, a play-within-a-play. There are several other theatrical insets in Kyd's play, which leaves no doubt that the whole has a conspicuously theatrical construction, that is, he uses theatre conventions not only to construct a play, but also to speak metaphorically about the outer world.

Revenge is not the only one who stages and interprets the play (actually, several different spectacles). In act 1, scene 4 it is Hieronimo, and not Revenge, who presents a dumb show that he has devised:

Enter Hieronimo with a Drum, three Knights, each with scutcheon: then he fetches three Kings, they take their crowns and them captive.

This scene parallels the ones involving Revenge and Andrea: a theatrical spectacle is presented and is commented upon through a dialogue between a naive spectator (i.e. the King) and the author or critic (i.e. Hieronimo) who interprets the visual, non-verbal side of the performance. Immediately, we may notice the difference in the time structures involved. The difference, however, remains in that the first dumb show may be treated as a playlet performed before two spectators only in the play without, whereas the second one is also perceived by the Spanish court. From the point of view of the Royal family and the courtiers, it is simply a masque staged at court; from the point of view of the Ghost and Revenge it is not a play-within-a-play, but rather a real life situation, in which real people watch a masque; from the point of view of real spectators, both situations are examples of a play-within-a-play. In both cases equally important to what is being presented in the show within are the reactions of the spectators in the play without. This also shows an interplay of several semiotic orders, each of which is governed by a different hierarchy of functions.

Crucial, of course, is the playlet within in act 4, which is part of Hieronimo's scenario. This time, it is neither a masque nor a dumb show, but a proper play, with dialogues, action and so on. So, the hiatus between the temporal and spatial structures of the framing inner play and the play within are clearly marked, and so is the new *deixis* of the figures within whose "I" differs from the "I" of those who enact them, even though the amateur actors are the figures from the former sphere. In accordance with the rules of the theatre, the present time of the figures in the inner micro-play is different from the present time of its spectators, inhabiting the world of the framing inner play.

Please note that in the actual staging of Hieronimo's play, the different languages employed are also recognized as such by the stage figures in the macro-play.[16] It does not mean, however, that they we are dealing with heteroglossia in the play within, i.e. that the figures in the play of *Soliman and Perseda* speak in sundry languages. The playlet is followed by Hieronimo's oration, in which he explains the true significance of the presented scene; as he has promised, he also employs yet another theatrical device: his son's corpse is hidden behind a curtain, and a stage prop is used, a bloody handkerchief. In this particular case he reveals the fact that the boundary dividing empirical reality and fictional reality on stage has been abolished (within the fictional realm created on the stage). The death of the stage-figures means also the death of the actors and, from the point of view of the Ghost and Revenge, of the "real people". That violates the basic rule of theatre, which is the separation of the actor from the figure he/she enacts. The actions of the figures cannot be transmitted to the ontology of real life actors, so if someone gets hurt on the stage it is the figure, not the actor. It is the figure that dies, cries, laughs, sleeps, and not the actor. If the actions of the figure are transmitted on to the physicality of the actor then they merge completely; the fictional present time evaporates and is replaced by the real biological time of the actor. They find themselves in one cause and effect sequence. At this point theatre stops being itself, and is to be replaced by some other form of human behaviour. This is exactly what Hieronimo plans to do: he stages the play in order to solve his problems in the real world. According to his scenario, the end of the play will be marked by the end of Hieronimo's life. Human life is literally incorporated into play reality, as in Roman amphitheatres where actors were killed for real. Again, the interpretation of the play within depends on the point of view: it is different for the two spectators on the stage, for whom the actors really die, and different for us, the real spectators, for whom both spheres are fictional.

Hieronimo's scenario goes astray; it fails on at least three counts. First of all, Bel-imperia commits suicide. Secondly, Hieronimo is prevented from hanging himself; and thirdly, in his fury and madness he kills the Duke of Castile – something not even considered in the scenario. Naturally, things go wrong only on the level of the play within. In the macro-play, everything goes in congruity with Revenge's script, and he ends the play with the following:

> Then haste we down to meet thy friends and foes,
> To place the friends in ease, the rest in woes:
> For here though death hath end their misery,
> I'll there begin their endless tragedy.

> (4.5.45–8)

Thus we return to the initial opposition of "here" and "there". "Here" does not necessarily means "theatre" in the architectural sense, but, more precisely, the place from which the Ghost and Revenge observed humans at action; "there" stand for the metaphysical sphere, the underworld. Symbolically, this could mean transitoriness versus eternity. Death in one sphere marks the beginning of eternal joy or suffering in the other. The end of life is metaphorically the end of the play, but earthly tragedy marks the beginning of the endless one. The figures from the Spanish court transgress the boundary that divides life and death, metaphorically represented in the structure of the play. So, it seems that at last we see the two distinct streams of time merge: the end of the play is the beginning of the metaphorical tragedy, which is not shown on stage. The title of Kyd's play refers to the latter; the Spanish tragedy is the final act, which, surprisingly, is not enacted, as if Kyd had forgotten about something (the play has four acts only). But Kyd was an experienced playwright, and the missing act is not an error, but a conscious act of signification. The King is dead, so are his possible successors; there can be no greater tragedy for a state, according to Renaissance political thought.[17] This explains why the play has only four acts: the fifth one is only implied, and is "endless". Indeed, the "earthly" part of the play has to be left as if "unfinished", but its four acts lead us to the final one which is taking place in the underworld and which has continued ever since Kyd wrote his play, and will continue for ever. The implication is that what we have been watching was not a theatrical presentation of events that have already passed, but the real stream of time in which people live, and that time was eventually "catching up" with the real time (timelessness?) of Andrea and Revenge, to merge at the very end of the play. Please note that the earthly time merges with the metaphysical one and for that reason cannot have an end.

Conclusion

The chemical metaphor used throughout the book derives from my under-standing of theatre as art, in which theatre is seen not only as whatever occurs on the stage, but as a complicated communication process, by which a sort of feed-back is established between the phenomenology of the stage, the denoted fictional realm and the mind of the spectator. These were described as chemi-cal retorts, in which all sorts of reactions take place, and which interact with one another as if in communication vessels, culminating in cognitive and aesthetic pleasure. It is only through the reactions between the three retorts that theatrical meaning is established. This metaphor enabled the author to make an attempt to define theatre, which today, when the demarcation lines between different arts and life have become blurred if not altogether erased, may be regarded as a somewhat risky undertaking. Especially in the present day, when, following the example of art, we are silenced by strict and omni-present demands that we should give up the traditional humanist longing for clearly defined categories. We are told that definitions and demarcations are suspect and reactionary; so is theory. There is no doubt that performance art, body art, video art and the like have entered the theatre on a large scale during the last two decades or so, and some aspects of theatricality are clearly noticeable in the unbelievably complex and varied forms of what art is today. But this does not mean that we cannot distinguish between the particular artistic disciplines, and it seems, as this book attempts to show, that time may be the key factor in enabling us to make specific distinctions. Moreover, the interaction between art, new media and theatre can be extremely fruitful, as for instance The Wooster Group has shown us, but, again, this does not mean that we have to treat them as one artistic phenomenon.

Among the chemical metaphors I have employed, time was presented as the binder of all things theatrical. And, indeed, time is the major theme of this book. I am aware that the complexity of theatre chemistry and the human mind is too great to cover in a single book. There are many areas of investiga-tion which have been barely touched upon, or even entirely neglected. What I have intended to show, however, is that the study of the different functions of time structures employed in theatre is fundamental in uncovering the ways in which the medium works, the complex and distinctive ways in which meaning is created. Already in the introductory chapter I defined theatre as a communicative situation, in which someone (the "actor") pretends before someone else (the "spectator") that what belongs to the latter's past, is in fact his/her (i.e. the actor's and the figure's that he/she enacts) present time. In doing this, the actor assumes the role of someone else, living at a different time and in a different space. In other words, events from the past (rarely from the future) are presented as if taking place in the here and now of the spectators. Of primary interest here are the time structures that create a unique set of relations. Consequently, in theatre we have two present times, one that is real, and includes the biological time of both the actor and the spectator, and one that is fictional, and belongs to the world of the figure. It should also be mentioned that the two present times belong to two different streams of time. The fictional time is finite; it has a marked beginning and a marked definite end, and does not extend beyond those points of demarcation. The latter fact explains why even those performances which seem to be "real", enacted by human beings rather than stage figures, in the here and now of spectators, and which for this reason present themselves as "a slice of life" (as in Heiner Mueller's *Offending the Audience* or in the so-called postdramatic theatre), are in fact conventions that try to hide the temporal distinction. Even in those performances where the liveness of the performers is stressed (or "foregrounded") the fictional present time is marked by the fact that it belongs to a stream of time that is iterative, preordained and clearly demarcated. And within the marked time of the performance, everything that appears on the stage undergoes semiosis and enters into closer or more distant relations (reactions) with everything else. All communicative situations which reveal the features mentioned above are theatrical in nature.

In this book, I have discussed all sorts of issues connected with theatre, and on each occasion time was the decisive factor that enabled me to clarify and show the theatricality of a particular issue. The study also includes areas of investigation that usually are not considered as based on the play of different time structures, such as acting or the language used on the stage. And yet, as seems to have become apparent in my discussion, time is the key factor that enables us to identify stage speech and, surprisingly, acting. The latter I have defined briefly as the ability to create fictional present time by someone who is a live person, i.e. living in the biological present shared with the spectator. Thus, throughout this book I have been trying to identify and clarify theoretical problems connected with the most complex and difficult of all artistic mediums, i.e. the theatre. In dealing with them I have employed analytical tools which treat time as the key component of the theatrical performance and the instrument in deciphering practically all the issues involved, starting with the effect of theatricality, through the complexities of impenetrable stage speech, to theatre conventions like soliloquies, asides or theatre-within-the-theatre. In the same way as time has been described as the binder of all things theatrical, time also holds this book together. It is the principal actor among the presented theorems.

I am fully aware how difficult a topic time is, both for philosophers and psychologists, but also for all those who attempt to theorize on issues linked to the arts, literature and theatre. For this reason, I consider my book not as a definitive work, but as a small introductory step towards further, no doubt deeper and more thorough investigations. Having finished the work, I have fewer doubts than ever before that it is indeed time that opens the doors to areas of research that can give answers to seemingly insoluble questions. And especially so today, when the humanities are tarnished by an ongoing ideological struggle, which may end with the victory of those who wish to abolish all boundaries, who claim that theory is a reactionary and fossilized area of research, and that any attempts to define basic notions, categories, and areas of investigation are bound to fail. What follows is the claim that the boundaries between the arts or theatre and life have finally become completely blurred or have disappeared altogether, and that in fact there is no difference between an artistic event and a social occasion. But I strongly believe that theatre theory can be a scholarly discipline, even a kind of science, if you will, and not just a set of arbitrary opinions. The intention is not to "jail" the theatre, but, on the contrary, to liberate it from the misunderstandings and confusion caused by ideological disputes and controversies. I have attempted to base the discussion of theatre on solid assumptions, verified by logic and reason, by the findings of other disciplines and, perhaps above all, by practice, which can prompt solutions that are tenable and equip us with appropriate analytical tools to ensure a sound discussion. It seems that time, with its various functions and implementations in artistic texts or events, may in fact provide an objective and verifiable criterion for making the distinctions that are now being questioned. It remains my hope that I have managed to set out the rudiments for the study of time, even though it encompasses a relatively narrow area of investigation. Even if the answers reached are not always satisfactory to all, I think Shakespeare would agree with me that it is our duty to defend our right to ask questions. In none of his works does he provide ready solutions to the questions asked, but he stresses the importance of asking them.

Notes

Preface

1 This is proved by the persistent use of the word "drama" to denote theatre as an artistic form, especially in the English-speaking world. Even Pauline Kiernan, who has spoken against many fossilized ideas current in theatre studies, entitled her book *Shakespeare's Theory of Drama*, and she uses the word ("drama") throughout in both a literary and a theatrical sense. Similarly, in a recent essay, Noël Carroll talks about "different arts of drama: the art of composition or creation, and the art of performance or execution"; see Noël Carroll, "Philosophy and Drama. Performance, Interpretation, and Intentionality", *Staging Philosophy. Intersections of Theater, Performance, and Philosophy*, ed. by David Krasner and David Z. Saltz (Ann Arbor: University of Michigan Press, 2006), p. 107. Jerzy Grotowski puts this in extreme and paradoxical terms: "what is good for the theatre is bad for the drama, and what we have customarily labelled as drama (the verbal structure created by the playwright) is always unsuitable for the theatre". Of course, one does not have to agree with Grotowski.

1   Reading the Elements and Compounds

1 The word "theatre" has several meanings: it may mean a building used for the presentation of stage plays, it may mean an institution (a company of actors), and it may also mean an artistic system or medium which generates meanings with the help or distinctive rules (like other systems such as literature, painting or music). In this book this latter meaning will be most frequently employed. I am not going to take metaphoric usages, very common today, into consideration. See discussion of the definitions in Samuel Weber's *Theatricality as Medium* (New York: Fordham University Press, 2004), pp. 97–103.

2 The category of truth is a philosophical one, and I have no intention of entering into its discussion here. Let me just quote Tennessee Williams, who, through his *alter ego* in *The Glass Menagerie* says: "a stage magician . . . gives you illusion that has the appearance of truth. I [i.e. the playwright] give you truth in the pleasant disguise of illusion".

3 The inverted commas around "language" are necessary here, because with reference to theatre we can talk about it only metaphorically. See Gay McAuley, *Space in Performance: Making Meaning in the Theatre* (Ann Arbor: The University of Michigan Press, 1999), pp. 6–7, and also my discussion below in Chapter 6.

4 In today's theatre we witness the growing role and importance of the dramaturg. See, for instance, Cathy Turner and Synne K. Behrndt, *Dramaturgy and Performance* (Basingstoke and New York: Palgrave Macmillan, 2008).

5 Cf. Eli Rozik, "Theatrical Convention: A Semiotic Approach", *Semiotica* 89.1/3 (1992), pp. 1–23. Characteristically, whenever a convention is employed for encoding the message, we grasp the fact of its being employed, and since it changes the mode of stage utterance, we understand that it is the author who models his text in a new way to convey more information. And the particular device used is the

best medium for conveying that information. It is not out of Hamlet's habit or personal need or will (he has none) that we hear him as if talking to himself: it is the playwright's employment of a known convention (i.e. the soliloquy, discussed in detail below) by which he is capable of conveying more information about the Prince and his perception of the world around him than in a "less-conventional" dialogue. We also notice that conventions are not perceived by the stage figures that treat the world around them as real, and not one governed by stage rules.

6 Of the numerous books on the subject see Erika Fischer-Lichte, *Estetytka performatywności* [*Ästhetik des Performativen*], translated by Mateusz Borowski and Małgorzata Sugiera (Kraków: Księgarnia Akademicka, 2008) and Marvin Carlson, *Performans*, translated by E. Kubikowska (Warsaw: PWN, 2006).

7 I am aware that the word "signifier" is not widely accepted in today's theoretical writings. I do not insist on using it, and what I mean is just the signalling matter of the stage, which is used to model the performance and to convey meaning.

8 We must not forget that the uniqueness of a production finds its variants in actual performances, which inevitably differ, at least slightly, one from another.

9 "Transmutation" is a term introduced to semiotics by Roman Jakobson, meaning intersemiotic translation, whereby in theatre various elements of the verbal text (not only the stage directions) are "translated" into actors' gestures, costumes, elements of the set, music, dance and so on. See Roman Jakobson's article on linguistic aspects of translation; I have used the Polish edition, "O językoznawczych aspektach przekładu", *W poszukiwaniu istoty języka*, ed. by Maria Renata Mayenowa, translated by L. Peszczołowska (Warsaw: PIW, 1989), vol. 1, pp. 372–81.

10 I agree with Eli Rozik's comment that: "In contrast to a widely accepted view, actors are not 'impersonating' . . . or 'pretending' . . . but honestly employ their crafts and the underlying rules of their medium to the best of their capabilities to describe characters, which is their professional task" – *Generating Theatre Meaning* (Brighton and Portland: Sussex Academic Press, 2008), p. 87.

11 The theoretical basis of theatre semiotics was the creation of Czech and Russian formalists. I am using the term "semiotization" to denote the process of adding sign function to objects, bodies and all material components of a performance. This explains why everything visible and audible on the stage has to be treated as a sign of something else that exists in the fictional realm. In other words, everything becomes a sign of what it is not. On semiotics in theatre studies see the most recent book by Eli Rozik, op. cit.

12 Spatial issues will be discussed in Chapter 4. Fictionality in the theatre depends above all on a considerable sensory inaccessibility of the world created on the stage or on the material inadequacy of the signifiers. From the point of view of the external observer – "that which is talked about is not what is seen, that which is seen is not what is being talked about", and the experienced recipient knows that the reception of this world is not possible exclusively through the senses but demands the arousal of those regions of our consciousness known as the imagination. And on this also depends the scenic synaesthesia. The word wants to influence sight, so that it can "see" what is not there.

13 There exist many definitions of the word "code"; cf. Patrice Pavis's dictionary of theatre or, for instance, Marco de Marinis, *The Semiotics of Performance* (Bloomington and Indianapolis: Indiana University Press, 1993), pp. 97–8.

14 The spectator can only notice the outcome of various reactions occurring in the reality of the stage, e.g. someone breaking a glass, but cannot see the meaning of this in the fictional realm. The glass that breaks on the stage does not exist

on the other side. In *Millenium Misteries* by Teatr Biuro Podróży (1999), the scene showing the Slaughter of the Innocents was presented by the actors who smashed watermelons on the stage.

15 See Soshana Avigal and Shlomith Rimmon-Kenan, "What do Brook's Bricks Mean?", *Poetics Today*, 3.3 (Spring 1981), pp. 11–34.

16 Gilles Fauconnier and Mark Turner, "Blending as a Central Problem of Grammar" (1998), an article available on the homepage of the authors. See also their book *The Way We Think. Conceptual Blending and the Mind's Hidden Complexities* (New York: Basic Books, 2002).

17 Pierre Jacob and Marc Jeannerod, *Ways of Seeing: The Scope and Limits of Visual Cognition* (Oxford: Oxford University Press, 2003), pp. xii–xvi.

18 See a stimulating article "Falsifiable Theories for Theatre and Performance Studies" by Bruce McConachie, *Theatre Journal*, 59.4 (2007), pp. 553–77.

19 Ibidem, p. 566.

20 Indexical relations are based on spatial and temporal contiguity, and/or on the cause and effect sequence of events. The term itself derives from the seminal work of Charles S. Peirce, who differentiates three types of sign: icons, indexes and symbols. The icon is a sign that refers to the object denoted by means of characteristics with which it is itself endowed (similarities), regardless of whether the given object really exists or not. An icon may possess these characteristics in itself, or they may be bestowed on it in a given artistic utterance (we speak then of attributed similarity – a phenomenon typical of the theatre). In the case of the index we have above all – if similarity is lacking – a relation of contiguity between the material of the sign and its meaning (dark rings under the eyes indicate a sleepless night or tears, or – by a surprise – a special cosmetic "mask"). This type of sign appears most frequently in the theatre in a relation based on ostension or – which in the case of acting is the most important feature – on signalling a cause–effect relation that results directly from spatial contiguity (as in the given example of dark rings under the eyes). The symbol, in contrast, creates meanings of a connotative nature; it has only the value attributed to it (e.g. by cultural codes, religion or mythology), so neither similarity nor contiguity is expected of it. Symbolically dark rings may signify an evil character or drug addiction. We should recall here that according to Peirce the ideal sign has a complex character and all types appear in it at the same time, though in varying proportions.

21 In my understanding, "aesthetic" does not mean "beautiful" in the everyday meaning of the word. What it means is the appearance of the poetic or aesthetic function, which means that the work "explains itself" by showing its structural features, by drawing the spectator's attention to the systemic rules, which make the appearance of the work in the "form" given justified. In order to achieve that, the work (artistic event) has to bring to light the rules that enabled its formation.

22 See *The Senses in Performance*, ed. by Sally Banes and André Lepecki (London and New York: Routledge, 2007).

23 In Westerns, to give a cinematic example, the particular tune always announces the approach of the Indians.

24 The first to draw the distinction between the actor and the figure was Otakar Zich in his book *Estetika dramatického umeni* (Praga, 1931), pp. 213–14.

25 It has to be added at this point that the new wave of the so-called postdramatic theatre often attempts at breaking the boundary dividing the actor and the figure; consequently, the temporal and spatial split or hiatus is annulled. That may, of course, be a temporary feature of a production, but in its dominating variety it

undermines the basic qualities of theatre as art, which inevitably thus becomes another type of art, such as performance or happening. Cf. Hans-Thies Lehmann, *Postdramatic Theatre*, trans. by Karen Jürs-Munby (London/New York: Routledge, 2006). However, as the discussion below shows, the apparent identity of the actor and the figure in postdramatic theatre is yet another convention: the performer is not the figure, because (1) he/she performs the same text every night before a different audience, and (2) even if his/her time overlaps with the time of the audience, the past and the future of the performer are not finite and equal to the demarcations of the performance, and for this reason they are different from the past and the future of the fictional figure.

26 It is plausible, of course, to have an actor, say, John Brown, impersonate himself, but this will be a fictional figure anyway, living at a time different from the time of the performance (John Brown as he was two weeks or five years before). If an actor plays him-/herself as living in the time of performance, this will not be theatre, but a different kind of show, a cabaret, for instance.

27 See Stanton B. Garner, Jr., *Bodied Spaces. Phenomenology and Performance in Contemporary Drama* (Ithaca and London: Cornell University Press, 1994), especially chapter 4 "The Performing 'I': Language and the Histrionics of Place", pp. 120–58.

28 Jacqeline Martin, *Voice in Modern Theatre* (London and New York: Routledge, 1991), pp. 138–40, Jon Whitmore, *Directing Postmodern Theater. Shaping Signification in Performance* (Ann Arbor: The University of Michigan Press, 1994), pp. 75–6.

29 It may only be added that in the postdramatic stream of the theatre, the plot, intrigue, suspense and the like often lose in their importance, so does the language, and what is foregrounded is the "pure" theatricality of the show, i.e. those elements that point to the material nature of objects and bodies that have been selected and to how they are being staged. Following the earlier revolution in art, also in theatre the basic concept of an artistic "work", or artefact, is rejected, and what comes to importance is the unique experience between the performer (the word actor is avoided) and the spectator, or, rather, participant Thus, we have entered into a new phase in culture, art and theatre, which is not only postmodern, but also postaesthetic and posthistrionic.

30 See the discussion of illusion in Eli Rozik, *Generating Meaning in Theatre*, op. cit., pp. 177–9. This is confirmed by cognitivist studies. As Bruce McConachie has put is, "cognitive theories suggest that spectators understand the world on the stage not as an illusion, but as a different kind of reality when they are living in the blend of performance and mirroring the actions of actor/characters or looking at the setting of a production" (op. cit., pp. 565–6).

31 I have described the concept of the *fifth wall* in my book, published in Polish, entitled *Piąty wymiar teatru* (Gdańsk: słowo obraz/terytoria, 2006). See also my article "The Fifth Wall: Words of Silence in Shakespeare's Soliloquies and Asides", *Shakespeare Jahrbuch*, 144 (2008), pp. 47–65.

32 Interestingly, cognitive psychology has investigated the perceptual moment and experience of the present (now). According to recent studies, the perceptual moment with an outer range of 2–3 seconds serves to link data into a coherent unity, which forms the basis of our concept of the present. There is additional evidence that music, poetry and language (possibly the theatre?) are also segmented into intervals of up to 2–3 seconds. See Vivyan Evans, *The Structure of Time. Language, meaning and temporal cognition* (Amsterdam/Philadephia: John Benjamins Publishing Company, 2004), pp. 26–7.

33 In theatre we may have all sorts of objects and phenomena (light, sounds) that may denote the flow and shifts of time, but signs of that sort gain meaning only when the temporal value is attributed to them (verbally or through gesture, gaze and the like) by a human agent.

34 Moreover, psychologists and linguists have argued that time as such is not perceived at all: what is perceived are the events occurring in three-dimensional space – thus, "the concept of time is constituted by virtue of the motion events which serve to define event-comparison" (Evans, op. cit., p. 15). This feature makes it possible in theatre to arrange motions and events in space in a way that may be strikingly different from our everyday experience and may defy the laws of physics, and by doing so, mark different temporality of particular events and relations between them, for "literal time is a matter of event comparison", as Lakoff and Johnson have stated. Writing a letter on the stage may last fifteen seconds, reading a newspaper article twenty-five, a battle will occupy not three days but three minutes, and the family dinner will take five, coffee and dessert included.

35 Sometimes different figures see different things on the stage, as, for instance, in scenes where a ghost appears to one figure and does not appear to another one (as in the closet scene in *Hamlet*, where the Prince sees the Ghost and his mother does not, or in 5.1 of *The Revenge of Bussy D'ambois*, where the Ghost appears to Clermont and not to Guise). A variation of the ghost scene is the one in *The Atheist's Tragedy*, where in 2.6 the Ghost appears first to the sleeping Charlemont in his dream, and although his presence on the stage is certain (confirmed by a stage-direction "*Enter the Ghost of Montferrers*"), he remains invisible to the Musketeer ("You dream/Sir; I saw nothing", ll. 44–45).

36 Eli Rozik in his article "Acting: The Quintessence of Theatricality" (*SubStance*, 31.2&3 (2002), p. 113) also notices this triad, but includes different components in it: the actor, seen as the creator of signs, the text or the "pictures" that are created on the surface of the actor's body, and the figure, which is a mental construct. Veltrusky, on the other hand (in "Contribution to the Semiotics of Acting", *Sound, Sign and Meaning. Quinquagenary of the Prague Linguistic Circle*, ed. by Ladislav Matejka (Ann Arbor: The University of Michigan Press, 1978), pp. 576–92), talks of three functions of acting: expressive, conative and referential.

37 Evans, op. cit., p. 25.

2   The Chemical Formulas of Blending

1 Bernard Beckerman, when discussing stage fights, makes the following comment: "An actor may be carried away by the heat of the performance or, in order to impart maximum reality to the fight, the actors may reduce the margin of safety and come close to doing actual harm to each other. When the audience picks up these signals, it no longer enjoys the fight, but finds it painful and distasteful. Shows of illusion demand that the acts remain illusory and in dialectic presentation that the resistance be virtual and not actual" (*Theatrical Presentation*, p. 59). This was written before the flood of literal presentations on European stages, before the appearance of body art or the "in your face" stream. However, the rule has not changed, but needs a minor modification. It may be said that the literal is "distasteful" because it violates the rule of accessibility: the two worlds, the phenomenal and the fictional are divided by the invisible fifth wall, and should not violate the border. The actor engaged in a fight signals fictional events occurring beyond the fifth wall; this is why he/she, as a live human being, cannot be

harmed by what is being presented. Fictional fists or knives cannot reach him/her. When he/she is harmed, it means that the fictional has crossed the boundary to become phenomenal. This means that theatricality is gone, and, instead, when the literal dominates, we are dealing with a different type of performance, or an accident. Of course, theatre as a medium (event, if you like) is very tolerant and an occasional breach of the division between the fictive and the phenomenal is widely accepted.

2   It is the actor who speaks verse, whereas the fictional figure speaks in prose (unless otherwise stated, which is very rare). Ariane Mnouchkine once observed that an actor walks differently on the stage, depending on whether he/she uses verse or prose. We may only add that the meanings of the words used are different, depending on the way and style, in which the actor walks on the stage.

3   As, for instance, in scene 1 act 4 of *As You Like It*, where Jacques notices the verse with which Orlando, as a true poetical lover, makes his welcome to Rosalind: "Nay, then, God buy you, an [if] you talk in blank verse" (ll. 30–31).

4   It must be noticed, however, that the language on the stage ceases to be just a verbal activity: it becomes something that I label stage speech.

5   For an intriguing discussion of theatricality see, among others, Eli Rozik, "Is the Notion of 'Theatricality' Void?", *Gestos*, 15.30 (November 2000), pp. 11–30 and "Acting: The Quintessence of Theatricality", *SubStance*, 31.2 & 3 (2002), pp. 110–24. My understanding of theatrically is slightly different, if not complementary to Rozik's insights. See also Erika Fischer-Lichte, "Introduction: Theatricality: A Key Concept in Theatre and Cultural Studies", *Theatre Research International*, 20.2, pp. 85–9 and Weber, *Theatricality as Medium*, op. cit.

6   As Anne Ubersfeld has put it, "Performance mediates between two different historical referents to which, in another era such as ours, a third is added, that of today's performance. The stage space that is represented brings together several referential fields between which it sets up a whole series of complex mediations" – Anne Ubersfeld, *Reading Theatre*, translated by Frank Collins, ed. and with a foreword by Paul Perron and Patrick Debbèche (Toronto, Buffalo and London: University of Toronto Press, 1999), p. 102.

7   This is where I do not fully agree with Eli Rozik, who repeatedly insists that the description of the fictional world is the essence of the stage text.

8   When I label the signalling matter on the stage inadequate, I mean this only in relation to the physical characteristics of the world outside theatre, known to the spectator from experience. From the point of view of the rules and artistry of the theatre, its substance, in most cases, is perfectly adequate.

9   Ironically, it would take some effort on the side of the actors to signal that the crown is not real, but made of paper, or that the flowers on the table are really fake, and not real. Spectators are so accustomed to the fact that what they see as inadequate or tacky or fake is real in the fictional realm that special information is needed to show them that what they see is the same as what it denotes in the fictional realm. But even then, i.e. when the iconic similarity may be assumed as absolute, the fake flowers in the fictional world will not be part of a stage set.

10 It may be noted in passing that in Elizabethan English "nothing" is a sexual innuendo: the men have "something" and the women have "nothing" ("thing" as opposed to "no-thing"). A well-known example comes from *Hamlet*, in the scene just before the play-within-the play is to begin.

11 See Marvin Carlson, "The Haunted Stage: Recycling and Reception in the Theatre", *Theatre Survey*, 35.1 (Spring 1994), pp. 5–18.

12   For a general discussion of stage properties see Andrew Sofer, *The Stage Life of Props* (Ann Arbor: The University of Michigan Press, 2003).
13   For a more detailed description of that production see my *Dangerous Matter. English Drama and Politics in 1623/24* (Cambridge: CUP, 1986), pp. 98–129.
14   J. R. Green, *Theatre in Ancient Greek Society* (London and New York: Routledge, 1994), p. 57
15   And conversely, when, for instance, Viola in *Twelfth Night*, describes her speech as "poetical", the speech itself is in prose (1.5.141–160).
16   See Limon, "The Fifth Wall . . . ", op. cit.
17   For ideologically engaged texts what really counts is the meaning and not the ways of creating it. This is why political correctness is an ideological criterion, not an aesthetic one. In the way art in general and theatre in particular are discussed today, one may be left with an impression that many authors forget that we are dealing with artistic texts. By rejecting the possibility of distinguishing art from non-art, theatre from non-theatre, we lose grounds for a theoretical discussion.
18   It may be noted in passing that by mere coincidence, the festival sponsor of this particular production was International Paper.
19   See, for instance, Stanton B. Garner, Jr., *Bodied Spaces. Phenomenology and Performance in Contemporary Drama* (Ithaca and London: Cornell University Press, 1994), chapter 4: "The Performing 'I': Language and the Histrionics of Place": 120–58.
20   See also Vimala Herman, *Dramatic Discourse. Dialogue as interaction in plays* (London and New York: Routledge, 1995), pp. 37–75.
21   For an intriguing discussion of the ways in which we comprehend time, see Evans, op. cit.
22   I am not the first one to notice this. See, for instance, Anne Ubersfeld, who writes: "we can say that the [stage] text has constructed its own referent upon the stage, and the stage space presents the text's referential space. The stage sign has the paradoxical twofold status of signifier and referent" (op. cit., p. 102).
23   "In general, in order to understand a performance-text there is no need to learn the theatre medium because, due to the typical motivation of iconic units by similarity, meaning is generated by what can be termed 'natural inference'. There are iconic units, however, that are mediated by a stage convention and, at least partly, unmotivated" – Rozik, *Generating Meaning*, op. cit., p. 32.
24   Naturally, the meanings generated by the production may be confronted with the traditional or singular reading of the Shakespeare play (say, by A. C. Bradley or by Jan Kott), and may in fact provide material for a fascinating analysis, but any incongruities noticed by spectators should not be interpreted as outrageous distortions of the classic.
25   On the specific nature of reading drama see Herbert Grabes's article "Staging plays in the theatre of the mind", *Reading Plays. Interpretation and Reception*, edited by Hanna Scolnicov & Peter Holland (Cambridge, CUP: 1991), pp. 94–109. See also Rozik, *Generating Meaning in Theatre*, op. cit., pp. 90–101.
26   McConachie, op. cit. (2007), p. 565.
27   Fauconnier and Turner, op. cit. (2002), p. 47.

3   Chemical Reactions, or Blending the Components

1   In point of fact, poetry is usually not recognized in the fictional realm (the poetic function appears only to the spectators). Romeo's and Juliet's first dialogue takes a form of a sonnet, and its meanings are determined by the situational context,

both verbal and non-verbal. The young lovers do not know how to talk naturally about love, they do not know how to express their true feelings, and their shyness is reflected in their use of the most conventional and therefore "artificial" mode of expression of that period, which takes the form of a sonnet reflecting the chemistry between them. The "form" is the meaning, and there is no way in which we can separate the two. Please note that Shakespeare's use of the sonnet form in this particular scene conveys meanings that would be absent if any other form of dialogue was used. And the final "shape" of the dialogue cannot be translated into any other without a significant loss or distortion (impoverishment) of the meaning.

2   This is why I cannot fully agree with Eli Rozik who maintains in practically all his works that a theatrical performance is basically a "description" of the fictional realm. Naturally, the stage text does describe or denote fiction, but that is not an end in itself, because what is equally important it the relationship between that world and the substance and modelling of the signifiers. If we want to treat theatre as an artistic medium, then inevitably we must notice that in it the aesthetic function comes into prominence. Therefore, theatre is not only about creating or describing fiction, but also about the rules that enable a particular production take the shape it was given, both in the selection of substances and in their combination or modelling. What counts is the relationship between the denoted meaning in the fictional realm and its material signifier.

3   Acting provides a very difficult theoretical problem, dealt with in Chapter 7. See, for instance, Jiří Veltruský, "Contribution to the Semiotics of Acting", *Sound, Sign and Meaning. Quinquagenary of the Prague Linguistic Circle*, ed. by Ladislav Matejka (Ann Arbor: University of Michigan Press, 1978), pp. 553–606, Robert Gordon, *The Purpose of Playing. Modern Acting Theories in Perspective* (Ann Arbour: The University of Michigan Press, 2006).

4   This in fact was the metaphor created in one of the court masques staged for the celebrations of the marriage of Princess Elizabeth (James I's daughter) and Frederick V of the Rhenish Palatinate in 1613. Cf. Jerzy Limon, *The Masque of Stuart Culture* (Newark, Del.: The University of Delaware Press, 1990), pp. 157–69.

5   This part of the chapter is partly based on my article "Shakespeare the Semiotician, or, Shakespeare Writes about His Own Art", in *Playing Games with Shakespeare*, ed. Olga Kubińska and Ewa Nawrocka (Gdańsk: Theatrum Gedanense Foundation, 2005), pp. 107–19.

6   Note that the dog is not a sign of a human spectator, but of itself as the spectator. However, the assigned role surpasses the dog's understanding.

7   Kurt Schlueter, "Introduction to TGV", *The New Cambridge Shakespeare* (Cambridge: CUP, 1990), p. 15.

8   Burt O. States, *Great Reckonings in Little Rooms. On the Phenomenology of Theater* (Berkeley, Calif.: University of California Press, 1987), p. 33.

9   In other words, Launce as the signifier becomes the dog in its past time, whereas Crab plays the dog-spectator in the present. So, what we have is an unusual situation, in which the "real" dog watches its fictional self in the past. However, the dog is not real, because it is also (unknowingly) a sign of a fictional dog in the time and space of the play. The seeming unity of time and space between the performers and spectators is always a side effect of the play-within-the play. The effect is caused by the fact that the point of reference seems to be the same (a performance within), but it remains "seeming" because the world presented to us on the stage (the macroplay) is always our past or future (and cannot be our present). In other words, during the play-within-a-play, the figures-spectators

(time level 2) and the real spectators (time level 3) are watching events occurring on time level 1. Since outside theatre it is highly improbable that any physical phenomenon will be observed simultaneously from different time levels, the sense of temporal unity comes as a natural solution to inevitable contradictions. See discussion of play-within-the-play in Part III.

10 On the surface it would seem that the dog behaves like a true, fictional stage figure – it does not recognize the fictionality of the world around it. Similarly, Launce does not reveal any awareness that he is a figure and is taking part in a performance. Of course, the actor who impersonates Launce has that awareness, but he behaves as if he is pretending to treat the world around him as real. So, the difference is that in point of fact we do not see Launce on the stage, but an actor impersonating Launce, whereas the dog we see on the stage is not impersonating his fictional self, but is given sign-significance from outside (by Launce), and consequently does not have any awareness of what is going on.

11 Again, I cannot agree with Burt O. States who claims that in this scene the dog "is simply being itself" (op. cit., p. 34). Although a bit further on States contradicts himself and observes that "A dog on stage is certainly an object . . . ; but the act of theatricalizing it – putting it into an intentional space – neutralizes its objectivity and claims it as a *likeness* of a dog" (ibid., pp. 34–5). This is much closer to what I am saying in this chapter. Naturally, live animals appear in other plays, not only by Shakespeare, as, for instance, in Peter Handke's *My Foot, My Tutor* there is a live cat. Again, even if animals cannot act in an artistic sense, they enter into relationships with the remaining components of the performance, and through this generate additional meanings (also because of their ability to break the illusion, and inability to create fictional time structures).

12 This apparently is not possible, since in his cast Launce does not have even one actor to impersonate him. Since he wants the dog to be the spectator, his only option is to play himself (from his past).

13 This is exactly what happens in John Arden's play, *The Happy Haven*, where we have several important scenes in which a dog appears, and a number of figures deal with it, but it remains invisible to the spectators (although it barks occasionally). As we learn from the "Author's Note", "The Dog is to be imagined by the audience, but must be understood to be seen by all characters on the stage – it is not a 'delusion'". Quoted from John Arden, *Three Plays* (Harmondsworth: Penguin, 1964), p.193.

14 "Pretends" because when Crab is expected to play the spectator, he is watching the events occurring within the same time span as the event observed by the real spectators. Unity of time in theatre often results in the sensation of oneness.

15 The fictional nature of theatre-evoked realms has often been compared to dreams, and numerous examples of the simile may be found in Shakespeare's texts.

16 See Patricia Dorval, "Towards a Stylistics of the Modes of Ostension", *Theatre Research International*, 18.3 (1993), pp. 206–14.

17 See, for instance, Francis Reid, *The Stage Lighting Handbook* (London: A&C Black, 2001), pp. 3–9. For a historical outline of stage lighting and its artistic use in today's theatre, see also Christopher Baugh, *Theatre, Performance and Technology. The Development of Scenography in the Twentieth Century* (Basingstoke and New York: Palgrave Macmillan, 2005), chapters 6 and 7 in particular.

18 It is worth pointing out that the spectator's memory is something altogether different from the memory that serves to record the real world. The past time in theatre cannot be identified with the individual memory of the past as such, but rather it appears as an element of the "stretched present" of the occurrences on the stage,

the work or event itself, which is a closed whole. The past of the work does not concern anything that had occurred before its beginning. What we are dealing here with is partly a temporal sign, which does not reveal all its features simultaneously, but demands time for doing that. Here lies one of many differences between theatre and life, something that some contemporary theories tend to forget. In other words, theatre as a work of art demands from the spectator the division of his/her memory into two "kinds": one that deals with the performance text, and two, the memory that encompasses the whole experience and knowledge of the spectator, making the cognition, blending and decoding possible. The first of these memories concentrates on the signals emanating from the stage and on the relationships between them, and also allows the temporal and special demarcation of the work. This enables us to treat the work as a separate structure, as something different form the non-structured world around us. Otherwise, it would not be possible to distinguish theatre from real life, and for that reason it would not constitute a coherent artistic structure.

19 Bernard Beckermann, *Theatrical Presentation. Performer, Audience and Act* (New York and London: Routledge, 1990), p. 34.

20 See, for instance, Fernando de Toro, *Theatre Semiotics. Text and Staging in Modern Theatre* (Toronto and Buffalo: University of Toronto Press, 1995), pp. 70–1.

21 Please note that the "bundle of sticks" mentioned in the example, may gain a symbolic or connotational significance, if the emblematic code is evoked (in the emblematic tradition this means "power", fortitude, ability to survive, for individual sticks may be broken easily, whereas a bundle cannot).

22 Some other examples are given in Sofer, op. cit., pp. 2–3.

4 Sculpting the Space, or the Retorts at Play

1 Interestingly, a literal presentation of a doll's house on the stage may be employed to achieve an artistic effect, as was the case with Lee Breuer's 2003 production of Ibsen's *A Doll's House*. As described by Martin Puchner, "In a remarkable performance by Maud Mitchell, the dolled-up Nora moves and speaks like a manic puppet, her fast-talking baby voice amplified by an extraordinary sound system and design. The entire production takes place on a miniature set, with toy chairs, toy beds, toy doors, toy pianos, toy everything, forcing the life-sized actors into the world of dead and animated puppets. An army of dolls is employed to reanimate this overly familiar play and in the process make it seem strange and unfamiliar. [. . .] The most notorious element of this production is that all the male characters are played by actors who are approximately four feet tall" – HotReview.org, visited on 10 March, 2008.

2 See, for instance, Greg Giesekam, *Staging the Screen. The Use of Film and Video in Theatre* (Basingstoke and New York: Palgrave Macmillan, 2007) or *The Potentials of Spaces. The History and Practice of Scenography and Performance*, ed. by Alison Oddey & Christine White (Bristol: Intellect Books, 2006), and Steve Dixon, *Digital Performance. A History of New Media in Theater, Dance, Performance Art, and Installation* (Cambridge, Mass., London: MIT Press, 2007). See also *Theatrical Blends*, ed. by Jerzy Limon and Agnieszka Żukowska (Gdańsk: Theatrum Gedanense Foundation, 2010).

3 As in a play by a Polish poet/playwright, Miron Białoszewski, entitled *Lepy* – "Fly-papers".

4 Gay McAuley distinguishes three categories of space in theatre: *stage space*, or the physical space of the theatre, *presentational space*, or what is presented and

displayed within that space, and the *fictional space*, which includes the implied fictional space; cf. *Space in Performance. Making Meaning in the Theatre* (Ann Arbor: University of Michigan Press, 1999), p. 29. See also Bence Nanay, "Perception, Action, and Identification in the Theater", in: *Staging Philosophy. Intersections of Theater, Performance, and Philosophy*, ed. by David Krasner and David Z. Saltz (Ann Arbor: University of Michigan Press, 2006), pp. 244–54.

5 See, for instance, R. Strong, *Britannia Triumphans. Inigo Jones, Rubens and Whitehall Palace* (London: Thames & Hudson, 1980), O. Millar, *Rubens: The Whitehall Ceiling* (London: Oxford University Press, 1958).

6 See, for instance, a thorough study of space and time in Anne Uberseld's *Reading Theatre*, op. cit., chapters IV and V, pp. 94–157 in particular.

7 Naturally, in today's theatre practice quite often no respect is paid to the ideology or meaning of theatre interiors, and we might have an agit-prop play shown in a church.

8 On the relationship between the development of lighting technology and theatre see Baugh, op. cit., pp. 94–144.

9 This part of the chapter is partly based on my article "Boska perspektywa: czas i przestrzeń w dramacie liturgicznym", in: *Paradoksy humanistyki. Księga pamiątkowa ku czci Profesora Andrzeja Zgorzelskiego*, ed. by Olga Kubińska and David Malcolm (Gdańsk: Wydawnictwo Uniwersytetu Gdańskiego, 2001), pp. 243–69. The bibliography of liturgical drama is very rich. I can recommend the list provided in *The Theatre of Medieval Europe. New research in Early Drama*, ed. by E. Simon (Cambridge: CUP, 1991).

10 Elizabethan theatres have been interpreted as sophisticated and complex spatial models of Renaissance philosophical thought, but all these approaches neglect the fact that the playhouses were not even designed by architects, but by carpenters, whose knowledge of Ficino or Bruno were meagre, to say the least. See my discussion of this issue in "From Liturgy to the Globe: the Changing Concept of Space", *Shakespeare Survey* 52 (1999), pp. 46–53.

11 Hence the Elizabethan drama's characteristic large segmentation of the text into short scenes, and also sequences of short scenes, each of which takes place in a different time and place. Presentation of these plays would be technically impossible if every individual scene were indicated by a change of scenery. Consequently, in today's theatres, directors face a considerable problem.

12 If even one is left, the stage retains its former meaning. This sometimes happens in the theatre by an oversight and always gives rise to a spatio-temporal contradiction in the following scene, and as a result a strong epic signal.

13 See also Masiko Ichikawa, "What to do with a Corpse?: Physical Reality and the Fictional World in the Shakespearean Theatre", *Theatre Research International*, 29.3 (2004), pp. 201–15.

14 There are exceptions from this rule, for instance when the stage is divided into different locations. If someone falls asleep on the stage, then it is rather difficult to find a good reason for taking him or her out. So, in a scene when X is shown on the stage asleep, and Y enters without noticing the sleeper, then we are led to believe that they are indeed in different places, not necessarily close by. So, when towards the end of scene 2 act 2 of *King Lear*, Kent imprisoned in stocks falls asleep, Edgar enters and presents a soliloquy, which as an exile and outcast implies that he can be anywhere but not in his father's castle (where Kent is). Where exactly Edgar is – is not explained. What we understand is that he is "somewhere else". This example shows also that the Elizabethan stage made use of the Medieval simultaneous stage tradition.

15 In *Coriolanus* (5.3), on the other hand, the scene takes place in what seems to be "inside the tent", which means that the whole stage becomes the floor in the denoted tent.

16 I have described this production in greater detail in an article "Space is Out of Joint: Experiencing Non-Euclidean Space in Theatre", *Theatre Research International*, 33.2 (2008), pp. 127–44.

17 Another example of the subjectivization of space on the stage is the 2006 London production of Christopher Marlowe's *Doctor Faustus* in Hampstead Theatre (directed by Rupert Goold): in the last scene, following a blackout, we see the two major actors (playing the Chapman brothers) and the stage set as if seen from "above". The question remains, of course, "who is looking", if the point of view is certainly not of any of the fictional figures and not of the audience (God?). The first version of this play, directed by Rupert Goold, opened in the Derngate Theatre in Northampton in 2004. Also, Veltruský gives several examples of subjectivization of space on the stage; for instance, in one Czech production as soon as the blind man approached the window, the vista turned into sheer darkness, stressed by the sharply lit window-frame (op. cit., p. 586).

5   Sculpting the Time, or the Magical Binder

1 Time is a very complicated issue, and has been tackled by various theorists and critics. Of the significant number of works let me just mention Patrice Pavis' *Dictionary of the Theatre: Terms, Concepts and Analysis*, translated by Christine Schantz (Toronto: University of Toronto Press, 1998), Richard Schechner, *Performance Theory* (New York and London: Routledge, revised edition 1988) and J. C. M. Pinto, *The Reading of Time. A Semantico-Semiotic Approach*, (Berlin and New York: Mouton de Gruyer,1989). Among numerous articles, the one that tries to make sense out of sundry approaches to time is Janette Dillon's "Performance Time: Suggestion for a Methodology of Analysis", *Medieval English Theatre*, 22 (2000), pp. 33–51. The present chapter is partly based on my article "Theatre's Fifth Dimension: Time and Fictionality", *Poetica*, 41, H. 1–2 (2009), pp. 33–54.

2 As Andrew Sofer brilliantly put it, "The director's job is not to realise all possibilities latent in the script, but to *sculpt stage time so that it moves meaningfully for an audience*" (op. cit., p. xiii).

3 Certain elements of this may be seen in children's play, when the participants enact assigned roles of, say, princesses and knights. This indeed is "theatrical". The difference, however, is that a game does not have an addressee other than the participants. Consequently, there is no discrepancy created between the time and space of the audience, and the time and space of stage figures. Without that discrepancy no theatre is possible.

4 This is not dependent on whether the ending is *open* and *closed*. The division relies on the plot, and not on the performance as a real event. An interesting example of an open ending may be found in Robert Wilson's *Threepenny Opera* (Berliner Ensemble 2007), where the play concludes with a "staging of a curtain". What happens is that when Mackie is miraculously saved from hanging by a sudden decree of the Queen, a bright crimson curtain falls, quite appropriate, aesthetically that is, to the trite and improbable ending. The real curtain, which opened the play, does not fall at all. The incongruity of two curtains brings about metaphoric interpretations. It may therefore be said that Wilson shows that plays with operatic endings are banal, and stresses that through the introduction of the

crimson curtain. However, the whole play as such, which raises issues that are important to humans, does not end, and therefore the real curtain does not fall.

5 Even in human language time is most frequently described in terms of space and movement ("the time has come", "a long time ago", "time is running out", "time has stopped", "to fall behind the time", etc.).

6 An example of the discussion of that kind may be found in Celia R. Daileader's *Eroticism on the Renaissance Stage. Transcendence, Desire, and the Limits of the Visible* (Cambridge: CUP, 1998), where, in chapter two, the author analyses the time of the offstage sexual intercourse on the basis of the action and dialogue occurring on the stage.

7 The metaphor of an hour-glass referring to the time of performance appears also in *Pericles*, where in 5.2 Gower says "Now our sands are almost run" (l. 1). Sand may of course become time's metaphor on the stage. Please note that "jumps over time", quoted above, cannot occur within a single scene, because otherwise it would be impossible to retain the duality of the present time, which is a continuum.

8 Of the numerous typologies of time in drama (not in theatre), let me mention just one, proposed by John C. Meagher in his *Shakespeare's Shakespeare. How the Plays Were Made* (New York: Continuum, 1997), pp. 69–94. Meagher enlists seven different times operating in drama: (1) natural time, (2) expanded time, (3) condensed time, (4) multiple time, (5) displaced time, (6) linked time and (7) intervening time.

9 Sometimes, objects may reveal this potential, but only in a very limited scope. For instance, lit candles on a dinner table, with no people sitting at it, with leftovers visible on the plates, glasses of wine half-empty, and a chair or two knocked over, could imply that some people left the dinner party hurriedly, not long ago. Since the candles are still burning, the time of the indeterminate event is implied, and so is also its continuation. It is the live flame of the candles that functions as a time marker, or time generating component. It reveals its ability to signal the fictional time; but it does that only because of the preserved indexical relation to whoever lit them. Once the flame is gone, the candles lose their ability to signal the continuity of fictional time. The same ability could be applied to running water in somebody's kitchen or bathroom, lit lights, or any other mechanical instrument that needs a human agent to start its operation (a TV set, a record player, etc.). It is not the objects out of themselves that can "enact" or signal fictional time, as Eli Rozik would like us to believe, but their indexical relation to an unseen human being that equips them with a temporary ability of that kind. See, for instance, Rozik's fascinating analysis of Robert Wilson's installation *H. G.* in *Generating Theatre Meaning* . . . (op. cit.), pp. 237–49.

10 Clocks are rarely met on the stage, for the simple reason that an operational clock, showing the time of the spectators, or even running with the same pace, would soon show temporal contractions, resulting from the different pace of the time shown by clock and the implied fictional time.

11 See Metin And, *Drama at the Crossroads. Turkish Performing Arts Link Past and Present, East and West* (Istanbul: Isis Press, 1991), pp. 119–20.

12 I have used this example in the article "Theatre's Fifth Dimension: Time and Fictionality", *Poetica*, 41, H. 1–2 (2009), pp. 33–54.

13 See Vyvyan Evans, op. cit., chapter 16: "Time, motion and agency", pp. 201–10.

14 The described phenomenon needs it, seems, further investigation (especially in the area of the psychology of time perception).

15 The following part of this chapter is based on my article, "Waltzing in Arcadia", published in *New Theatre Quarterly* (2008), pp. 222–8.

16   Some of the issues touched upon in this part of the chapter were dealt with, albeit differently, by Enoch Brater, "Playing for Time (and Playing with Time) in Tom Stoppard's *Arcadia*", *Comparative Drama* (June 2005), pp. 157–68

6   Sculpting the Language, or Stage Speech

1   Relatively few theoretical works deal with the question of the language in theatre. Relevant studies include those of Erika Fischer-Lichte, Eli Rozik, Marvin Carlson and Vimala Herman. Rozik's works in particular deserve attention; see his "The Vocabulary of Theatrical Language", *Assaph*, Section C, no. 2 (1985), pp. 15–26, "The Functions of Language in the Theatre", *Theatre Research International*, 18.3 (1993), pp. 104–14, and also "The Syntax of Theatrical Communication", *Assaph*, Section C, No. 3 (1986), pp. 43–57. The topic was exploited further by the author in his *The Roots of the Theatre* (2005) and *Generating Meaning in Theatre* (2008). An intriguing study of different languages at play on the stage is provided in Carlson's *Speaking in Tongues. Languages at Play in the Theatre* (Ann Arbor: The University of Michigan Press, 2006).

2   Sometimes the fictional language of the figures is not natural (as in glossolalia or in the theatre-within-the-theatre), but it has to be recognized as such within the fictional world.

3   Amy Cook, "Interplay: The Method and Potential of a Cognitive Scientific Approach to Theatre", *Theatre Journal*, 59.4 (2007), p. 589.

4   As observed by Roman Jakobson, quite contrary to its cinematic usage, language in theatre is not entirely an acoustic phenomenon, for it is also part of human actions (I have used a Polish edition: Roman Jakobson, "Czy upadek filmu?" in: *W poszukiwaniu istoty języka*, ed. by Maria Renata Mayenowa, vol. 1 ((Warsaw: PIW, 1989), p. 95). And these actions constitute an important part of the whole production. This partly explains the widespread popularity of speech-act theory in application to theatre studies. I shall return to this issue later.

5   Words enter into all sorts of relationships with other words, as is the case of literary or dramatic works. The difference in their scenic usage is, however, that they become inseparably blended with the human body and the material substance of a given performance.

6   And the "grammar" of theatre does not really exist. There are, however, some basic rules which any performance has to follow.

7   Almost a century ago the Prague semioticians observed that everything on the stage becomes "a sign of a sign"; what has been neglected from consideration is the language, which falls under the same rule. Consequently, all verbal utterances ought to be treated as signs of utterances within the created fictional realm, which by definition are not accessible to our senses. We hear the real actor-human being say something, but we do not hear King Lear say anything, simply because King Lear does not exist anywhere in the material sense. He may of course appear as a spectator's mental construct.

8   "Picture-like representations are variously described as having analogue, iconic, continuous, and referentially isomorphic properties, whereas language-like representations are characterised as being non-analogue, noniconic, digital or discrete (as opposed to continuous), referentially arbitrary, and prepositional" – Allan Paivio, *Mental Representations: A Dual Coding Approach* (New York: Oxford University Press, 1986), p. 16. For a discussion of related issues, see Howard

Mancing, "See the play, read the book", in: *Performance and Cognition. Theatre studies and the cognitive turn*, ed. by Bruce McConachie and F. Elizabeth Hart (London and New York: Routledge, 2007) pp. 189–206.

9   Jacob and Jeannerod, op. cit., pp. xii–xvi.

10 The language used by the actors is not natural, because it has already been modelled by the rules of drama. However, the implication is that fictional figures, who are not engaged in staging anything, use natural language. This is why the verse speech of Elizabethan plays becomes the prose of the denoted figures.

11 There are instances when words on the stage are not uttered by the actors, but are film or video recordings or are produced offstage. In those instances the relationship of words with the signalling matter of the stage will naturally be different. Please note that neither film nor video is simply "screened" on the stage: they are in fact staged. See Giesekam, op. cit.

12 For instance, in Chinese classical theatre, the drunkenness of the figure is not represented by an actor's movements or way of walking or talking, but by music (Veltruský (1984), p. 436). Thus, the music clarifies and determines the meaning of the actor's speech in a scene of drunkenness.

13 Cf. Elly Konijn, "Actor's Emotions Reconsidered: A psychological task-based perspective", *Acting (Re)Considered. A theoretical and practical guide*, ed. by Phillip B. Zarrilli (London and New York: Routledge, 2002), pp. 62–81.

14 See my article "The Fifth Wall . . . ", op. cit.

15 This was Michael Grunevitch's *A Fleeting Shadow*; see Eli Rozik, "A Silent Theatrical Production Representing a Speaking Fictional World: Analysis of 'A Fleeting Shadow' ", *Theatre Research International*, 27.1 (2002), pp. 78–96. Equally interesting in theatre is the ability of an actor to signal the presence of his/her interlocutor, without anyone else present on the stage. This is not necessarily a sign of the appearance of a ghost. For instance, in a seventeenth-century Spanish play, *The House of Desires* (RSC, 2004), an actor appears and from the utterances and gestures and movements of the body we understand that in the fictional world there is someone present, with whom the actor converses, but whom we do not see or hear. Naturally, we can have a reverse situation: we can see more than one actor/figure present, but they do not notice one another's presence. This means that they are separated by time and space, which is a common theatre convention.

16 In my approach to the function of language on the stage, I oppose many of the accepted concepts. For instance, in his analysis of speech acts on the stage, Eli Rozik comments: "it should be emphasised that the real referent of verbal categorisation is not stage reality but the fictional world evoked by the theatrical text: for example, 'stone' might be used for the categorisation of a lump of painted plaster and still be an adequate description for the fictional stone, if this lump is indeed meant to enact a fictional stone" ("Categorisation of Speech Acts in Play and Performance Analysis", *Journal of Dramatic Theory and Criticism*, VIII.1 (Fall 1993), pp. 118–19). Now, in my understanding the word "stone" refers to (a) the word stone as uttered in the language a fictional figure uses, and (b) to the "lump of painted plaster" on the stage. Both, the word and the lump are elements of stage speech. The meaning of this particular unit of stage speech is derived from the relationship between the word (and its meaning in the fictional realm) and the material substance of the stage stone. The word might mean something else in the fictional world than just a mere stone. It might be used in a metaphoric sense, as in "heart made of stone". Depending on that meaning, the semantics of the stage stone and the particular unit of stage speech change. Also, the meaning is partly determined by the substance

and its modelling of the stage stone, and by other material components of the unit of stage speech (music, the actor's body and gestures, light, etc.), as well as by the usage of the same word and the same object in other scenes.

17 Foreman's directing has been described in the following manner: "He sits in the front row of every performance, operating a tape recorder which often contains the entire text together with all the sound-effects, 'tape-loops', multi-vocal effects and music . . . This tape recorder is most important in Foreman's productions, as he records all the dialogue into it as a rule, so that in performance the actors only say a word or two quietly or repeat words as they are heard on the tape, echoing them slightly. Unless there is an accompanying gesture by one of the characters, the words heard remain anonymous – they are not related to any one in particular . . . He even instructs them [i.e. the actors] to deliver the lines flatly and in normal speaking voices. No attempt is made to project or interpret the lines, except occasionally when he decides to inject a moment of colour: then he gives a performer a specific instruction about how to interpret the line" – Jacqueline Martin, op. cit., pp. 138–9, 141. This implies that the voice that is heard on the stage has been almost completely detached from the body of the actor. In this way the voice itself becomes a constituent component of stage speech. This results is an unusual situation: the spatial contiguity between the word and the human body is undermined, if not annulled completely. It assumes, however, the quality of a radio voice, and may therefore be used as a component of stage speech. Robert Wilson, on the other hand, introduced, in some of his productions, "high-pitched screams . . . falsetto voice, slowed-down pronunciation, lapses into chanting in gibberish, disjoined phrasings and experiments with staccato rhythm, whispering and speaking 'past' each other" – Martin, op. cit., p. 147.

18 "We surmise that interchange of verbal elements within a fixed non-verbal context will not change the iconic sentence to any significant extent, whereas the interchange of non-verbal elements – with verbal ones remaining constant – will radically affect the meaning of the iconic sentence, including its verbal formulation. Thus, insofar as the receiver of the iconic sentence is concerned, the non-verbal component is subordinating in nature, as it determines the meaning of the verbal one" – Eli Rozik (1986), p. 50.

19 This is why the obscenity of the language in today's theatre practice may be interpreted as a sign of states of emotions rather than of the actual utterances of fictional figures. Acrobatics in Chinese opera may be a sign of a battle, and not only a demonstration of the acrobatic skills of the actors.

20 Rozik (1993), p. 120.

21 J. L. Austin, *How to Do Things with Words* (London: Oxford University Press, 1976), pp. 21–2.

22 One can agree with the commentary by Samuel Weber on this subject; see his *Theatricality as Medium* (New York: Fordham University Press, 2004), pp. 8–10. See also James Loxley, *Performativity* (London and New York, Routledge, 2007), pp. 62–87.

23 In my approach to speech act theory and its application to theatre studies I differ substantially from ideas expressed by many theatre scholars, even those of such renowned names as Eli Rozik (see his "Categorisation of Speech Acts in Play and Performance Analysis", *Journal of Dramatic Theory and Criticism*, VIII.1 (Fall 1993), pp. 117–32.).

24 Jiří Veltruský, "Contribution to the Semiotics of Acting", *Sound, Sign and Meaning. Quiquagenary of the Prague Linguistic Circle*, ed. by Ladislav Matejka (Ann Arbor: Michigan Slavic Contributions, 1978), p. 569.

25 On the other hand the existence of the referential function of the figure's language does not mean that the spectators will necessarily concentrate sufficiently and decode the meanings. More often than it might seem, audiences seem to concentrate on the fact of speaking rather than on the implied meanings, especially in set speeches of classical drama, which in today's world appear as tedious and too long, partly because the art of rhetoric is not recognized any longer. Veltruský, in another essay on acting, gives an example of an actor who forgot his lines at a crucial point of a play, and substituted them with glossolalia, nonsense verses, delivered with impassioned conviction; he was then applauded by the audience, with many of its members not noticing anything unusual in the speech. Cf. Jiří Veltruský, "Acting and Behaviour: A Study of the *Signans*", *Semiotics of Drama and Theatre. New Perspectives in the Theory of Drama and Theatre*, edited by Herta Schmid and Aloysius Van Kesteren (Amsterdam/Philadelphia: John Benjamins Publishing Company, 1984), p. 396 [393–441].

26 Marvin Carlson has devoted his entire book on this subject (*Speaking in Tongues*, op. cit.).

27 Limon, "The Fifth Wall . . . ", op. cit.

28 Michel de Certeau, "Vocal Utopias: Glossolalias", *Representations*, 56 (Fall 1996), pp. 29–47.

29 William Shakespeare, *All's Well That Ends Well*, The New Cambridge Shakespeare, ed. by Russell Frase (Cambridge: CUP, 1985), p. 112.

30 The case is different with staging foreign languages. See Andrew Fleck, "'Ick verstaw you niet': Performing Foreign Tongues on the Early Modern English Stage", *Medieval and Renaissance Drama in England*, vol. 20 (2007), pp. 204–21. And it is further different when several languages are used in one production, as is often the case in the postdramatic theatre. See, for instance, Carlson's *Speaking in Tongues* (op. cit.), especially chapter 4: "Postmodern Language Play", pp. 150–79. Again, also in the case of several languages employed, one has to distinguish between various possible situations: (1) the actors speak in different tongues, and so do the figures, with the result that they do not understand each other; (1a) the situation remains the same, but the figures understand each other; (2) the actors speak in different tongues, but the figures do not, and understand each other perfectly well; (3) the actors recite a well-known text in different translations, usually known to the spectators, and the figures understand each other perfectly well, as if one language was spoken only; (3a) the situation remains the same, but the figures do not understand each other (in a wonderful example that Carlson gives of a Shakespearean production, the figures try to understand the lines of their interlocutors by "translating" their gestures and mini-mime shows into the language of the original).

7 Sculpting the Body, or Embodied Time

1 Some of these publications carry learned titles, and may therefore be misleading. See, for instance, Rhonda Blair, *The Actor, Image, and Action. Acting and cognitive neuroscience* (London and New York: Routledge, 2008), which basically deals with actors' feelings and emotions.

2 Of the relatively few theoretical works dealing with acting, one ought to mention Patrice Pavis's *Analyzing Performance. Theatre, Dance and Film* (Ann Arbor: The University of Michigan Press, 2003), chapter 3: "The Actor", pp. 55–130, Eli Rozik's "Acting: The Quintessence of Theatricality", *SubStance*, 31.2&3 (2002),

pp. 110–24; see also Erika Fischer-Lichte, "The Actor's Activities as a Sign", in: *The Semiotics of Theater*, translated by Jeremy Gaines and Doris L. Jones (Bloomington and Indianapolis 1992), pp. 18–63, and Robert Gordon, op. cit.

3   See Erving Goffman, *The Presentation of Self in Everyday Life* (London: Penguin, 1959), and Richard Schechner, *Performance Theory* (New York and London: Routledge, 1977). Apart from the article mentioned in the note above, Rozik has dealt with the fallacy in his books *The Roots of the Theatre* and *Generating Meaning in Theatre*.

4   Hans Thies Lehmann, *Postdramatic Theatre* (London and New York: Routledge, 2006), and Erika Fischer-Lichte, *Estetyka performatywności* (Kraków: Księgarnia Akademicka, 2009).

5   See, for instance, Philip Auslander, *Liveness* (London and New York: Routledge, 1999), Simon Shepherd, *Theatre, Body and Pleasure* (London and New York: Routledge, 2006), *Performing the Body/Performing the Text*, ed. by Amelia Jones and Andrew Stephenson (London and New York: Routledge, 1999). It has to be stressed, however, that one cannot neglect the role of the body in the way we perceive and understand the world around us. This has become the focus of a recent trend in cognitive studies. See George Lakoff and Mark Johnson, *Philosophy in the Flesh. The Embodied Mind and Its Challenge to Western Thought* (New York: Basic Books, 1999), and Mark Johnson, *The Meaning of the Body. Aesthetics of Human Understanding* (Chicago & London: The University of Chicago Press, 2007).

6   I have published several articles on this topic in Polish journals: "Rozbieranie (się) teatru", in: *Inna Scena. Ciało, Płeć, Pożądanie. Tożsamość seksualna i tożsamość płci w polskim dramacie i teatrze*, ed. by Agata Adamiecka-Sitek and Dorota Buchwald (Warsaw: Instytut Teatralny, 2008), pp. 41–58, "Kneblowanie teatru", *Notatnik Teatralny*, 45–46 (2007), pp. 171–80, "Kto trzyma kaleidoskop, czyli o zalążkach terroru", *Dialog*, 10 (2006), pp. 100–11.

7   "Actors inscribe sequences of images of human behaviour, including speech, on their own bodies. These are indexical in nature . . . The theoretical problem is that indexical signs only refer to those who produce them, apparently the actors in the case of theatre . . . There is need, therefore, to explain how the actors who produce such indexlike predicates deflect reference from themselves to the characters who are supposed to produce them. They do so by enacting iconic signs in the capacity of both subject signs which identify a referent (the character) other than themselves and predicate signs which are eventually attributed to the enacted characters" – Eli Rozik, *The Roots of the Theatre*, op. cit., p. 21. This may, on the whole, be accepted as a valid statement. The only point worth raising now is that, as observed in the previous chapter, the actor's body is not the only object on which the language is "inscribed". Also, I have indicated that actor does all that "among other things", because, as this chapter attempts to prove, acting is much more than creating or describing a fictional figure.

8   Rozik (2002), p. 113. The distinction is further discussed by Rozik on pp. 114–17.

9   Gestures and movements have always been conventional in theatre, especially in the East. As to the European tradition, see, among others, Dene Barnett, *The Art of Gesture: The Practices and Principles of 18th Century Acting* (Heidelberg: Carl Winter Universitätsverlag, 1987).

10 Some of the examples of different functions of the actor's body may be found in Jeremy Lopez, "Imagining the Actor's Body on the Early Modern Stage", *Medieval and Renaissance Drama in England*, vol. 20 (2007), pp. 187–203.

11 There are instances, especially in the late twentieth-century theatre and performance art, of incorporating audiences into theatrical performances, without, however,

the spectators being fully aware of what is really happening. The confused spectators treat the pre-ordained events around them as real life, and are therefore not able to read them as theatre.

12 It is worth noting that David Z. Saltz has opposed some of these tendencies in his essay "What Theatrical Performance Is (Not): The Interpretation Fallacy", *The Journal of Aesthetics and Art Criticism*, 59.3 (Summer 2001), pp. 299–306. On the basis of what I have said so far, it may be added that since the actor is not the sole creator of the fictional figure, he/she cannot be held solely responsible for an interpretation of the dramatic "character".

13 See Eli Rozik's article "The Corporeality of the Actor's Body: The Boundaries of Theatre and the Limitations of Semiotic Methodology", *Theatre Research International*, 24.2 (1999), particularly his comments on p. 199.

14 Veltruský, op. cit. (1978), p. 558.

15 "Seemingly" because the image of the figure created in the mind of the spectator does not have to be the same as the one intended by the actor.

16 This is what Burt O. States calls the *self-expressive* mode of acting, the other two modes being *collaborative* and *representational*. See his "The Actor's Presence. Three Phenomenal Modes", *Acting (Re)Considered*, op. cit., pp. 24–8 [23–39].

17 See my article "Archaeology of Memory", *Cahiers Elisabethains* 73 (Spring 2008), pp. 39–47. Please note that when a film is screened on the stage, it ceases to be just a film: it becomes a staged film, and thus creates all sorts of relationships between its contents and the signalling matter of the performance.

18 Jiři Veltruský, "Contribution to the Semiotics of Acting", op. cit., pp. 574–5.

19 Rozik, "Acting: Quintessence of Theatricality", p. 115. Also, Rozik (2008, p. 84) extends the "enacting" function to practically everything on the stage, without taking into account that the ability of objects or animals to "act", i.e. their ability to deflect reference to fictional entities, rests entirely on their indexical relationship with a live actor. Without the actor they cannot "enact" anything.

20 See Elly A. Konijn, *Acting Emotions. Shaping Emotions on Stage* (Amsterdam: Amsterdam University Press, 2000).

21 See Robert Gordon, *The Purpose of Playing. Modern Acting Theories in Perspective* (Ann Arbor: The University of Michigan Press, 2006), pp. 221–58.

22 I have discussed Ciulli's production in greater detail in an article "A Candle of Darkness: Multiple Deixis in Roberto Ciulli's *King Lear*", *Journal of Dramatic Theory and Criticism* (Spring 2008), pp. 83–102.

23 The motif of the difficulty and even impossibility of communication is not the invention of the director. It is a development of what he read in Shakespeare's play. In *King Lear* the "positive" characters speak a highly individualized language, full of rhetorical figures, marked by poetic means of expression (with frequent apostrophes to the gods). But this language does not make communication any easier for them, indeed it frequently renders it impossible. It is difficult for them to communicate even among themselves, while the opposite side speaks a language of the concrete (with frequent apostrophes to nature), like *cliché* examples of politicians in negotiations or business people.

24 I shall give just one example: as is well known, the verse metre in Elizabethan drama best adapted to scenic dialogue is unrhymed iambic pentameter (so called *blank verse*). In *King Lear*, however, we find a departure from the norm: throughout the whole play, the Fool, played by the same actress who plays Cordelia, does not once use *blank verse*. This is by no means because he does not speak in verse: he sings rhyming songs, and yet he does not speak in *blank verse*, not even a single line. The

possible explanations of this are various, but the most likely is that the Fool speaks a language that does not invite dialogue: in his embittered wisdom, he has no desire to talk to anyone any more. The fact that he speaks results from his professional obligations, but also from deep disappointment with the world (and with his master), and not from his belief that his words can change anything, teach anyone or convince anyone of anything. Ciulli translates this into scenic images: The Fool's words fall into a vacuum, while the candle (a simple symbol of wisdom?) that he lights is not going to be seen by anyone (Lear does not notice it, despite his wide open eyes).

8   Soliloquies and Asides: Who's There?

1   Parts of this chapter are based on my article "Shakespeare's Soliloquies and Asides: A Theoretical Perspective", *Kwartalnik Neofilologiczny*, LII.4 (2005), pp. 301–23.
2   There seem to be differences as to the number of soliloquies and asides within particular dramatic genres: history plays will be less interested in the "psyche" of protagonists than tragedies, and will not employ varied dialogic devices for the sake of achieving comical effect, which often is the natural feature of comedies. So, for instance, in the whole of *King Richard II* we can find only one soliloquy, opening the final act. Asides, on the other hand, will be more frequent in plays involving intrigue and plotting, in all sorts of comedies of error.
3   Perhaps the most comprehensive definition and explanation of both a soliloquy and an aside may be found in Manfred Pfister's *The Theory and Analysis of Drama* (Cambridge: CUP, 1988), pp. 127–140. However, Pfister does not take into account the structures of time, which for me are vital in distinguishing between the two modes of speech.
4   James Hirsh in his *Shakespeare and the Theory of Soliloquies* (Madison: University of Delaware Press, 2003) insists on the distinction between a soliloquy, which is a sign of actual speech and an interior monologue, which is a sign of the speaking person's thoughts. However, Hirsch does not tell us how exactly are we to distinguish between the two, especially if a soliloquy is guarded from hearing of another figure present on stage. His distinction is based on psychological analysis, which does not recognize the fact that the language spoken on stage is not natural, and that the "psychology" of fictional stage figures is not the psychology of real human beings. In other words, the stage language is not the language which the figures use or the language they think in: it is only a sign of linguistic or mental activity, and taking it on the literal level betrays lack of understanding of theatre as an art form distinct from life.
5   This in fact is the basic premise behind James Hirsh's book, quoted above, in which the author attempts to prove that none of the Renaissance soliloquies were meant to be the sign of mental processes. On the contrary, Hirsh claims that speaking aloud was an accepted convention and each soliloquy delivered on the Elizabethan stage could in fact be overheard by one or more stage figures. Whenever he comes across a speech which is obviously meant to be a sign of actual silence, Hirsh dismisses it from consideration by labelling it an "interior monologue", which, according to him, is not a soliloquy.
6   Quite contrary to common belief, the soliloquy does not have to be uttered in the speaker's factual solitude: he or she may not be aware of other figures' presence; they might also be presented as if belonging to a different reality (i.e. a visual, but not necessarily mimetic, "illustration" of what is being said). I will go further and claim that a soliloquy may in fact be interwoven into the "normal" flow of a dialogue.

7   Sometimes (but not always) the fact of a figure speaking aloud to him/herself is clearly marked in the text. For instance, in *Antony and Cleopatra,* Enobarbus is talking to himself or rather praying to the moon aloud and three soldiers can hear him. It is a soliloquy since Enobarbus is not aware of anyone eavesdropping. It is therefore a sign of a true speech act within the fictional world and not of nonverbal mental processes.

8   Herbert H. Clark, Edward F. Schaefer, "Dealing with overhearers", in: Herbert H. Clark, ed., *Arenas of Language* (Chicago: University of Chicago Press, 1992), p. 256 [248–74].

9   The real audience of a given night is only the sign of an audience: this is a convention, and we should not treat soliloquies or asides as utterances of a real human being addressed to real people.

10   I am not the first to notice this; cf. Pfister, op. cit., p. 140.

11   Quite uncommon are the reasons for creating distance to (negative) figures by actors taking part in the traditional Turkish Ta'zije productions: they often make asides, directly to the spectators, and also comment on the cruelty of the figures they are creating. They do not do this in order to gain the sympathy of the spectators but to protect themselves from acts of aggression from them, since such "evil" figures are often exposed to such acts. Furthermore, they often keep in their breast-pockets pieces of paper with the text they are saying – they use this crib-sheet not because they cannot memorize their lines, but to show the spectators that someone else has written the script and that it does not come from the actors. In this manner distance and the destruction of the barrier separating the spectators from the performers becomes an element of the actors' self-defence. See Metin And, *Drama at the Crossroads. Turkish Performing Arts Link Past and Present, East and West* (Istanbul: Isis Press, 1991), p. 115.

12   Asides addressed to other figures are not taken into account here, because actually they are a specific form of dialogue. During their utterance, the fictional time flows with the same tempo, and there is no impression that the real actor was speaking. Also, addresses to other figures may be noticed by other figures, which is not the case with true asides.

13   At this point let me observe that although the present time seems to coalesce, the speaking "I" usually does not change. Also, what does not change is the stream of fictional time: this is proven by the fact that the utterance (i.e., the aside) is not spoken by the real actor out of him/herself, and also that the future time of the speaking "I" belongs to the finite time of the performance (the fictional stream) and not to the future of the real actor/human being and the future of the audience.

14   Pfister, op. cit., pp. 133–4.

15   Please note that this technique is often used in radio drama, with one figure speaking to another one whose answers we do not hear. In this way a dialogue takes the form of a dialogic monologue. Of course we understand that the technique is employed for particular reason: to create a monodrama, which will present the utterances of the speaking figure as actual speech acts and not his/her mental processes. So, all sorts of stylistic devices are employed, by which we understand that the speech is addressed to a present or imagined interlocutor, whose reactions are also recorded in the main stream of the utterance.

16   Cf. Pfister, op. cit., pp. 137–40.

17   Numerous examples may be found in Shakespeare's plays, as in *Richard III* in the scene with [old] Queen Margaret (1.3), or in *Cymbeline* (2.1), where the Second Gentleman makes six brief asides within forty lines of dialogue, and finishes the

scene with a typical soliloquy. All of the asides are directly linked to what is being said, and could match the dialogue perfectly well, if spoken aloud within the fictional reality. Also, in *The Winter's Tale*, in 4.4, Autolycus makes numerous asides during the ensuing dialogue, and all of his remarks are integrated with what the others are saying. Sometimes, as in the example of *Troilus and Cressida*, given above, it is not always certain whether a given aside is addressed to the audience or to one of the bystanders. The former would be a true aside, the latter – a form of a dialogue.

18 An earlier example of a dumb show inserted into a dialogue may be found in Thomas Kyd's *The Spanish Tragedy*.

19 I cannot agree with Bert O. States who claims that asides are typical for comedy and do not happen in tragedy, and when they do appear in the latter, they are utterances of fools or clowns or (clownish) villains. Cf. his *Great Reckonings in Little Rooms: On the Phenomenology of Theater* (Berkeley, Calif.: University of California Press, 1987), p. 170–5. All we may say is that in comedies metatheatrical devices are perhaps more frequent, and the aside – through its ability to unveil the conventions – has the effect of a sort of commentary on theatre as a conventional art.

20 That makes the play unique for Shakespeare: within seven consecutive scenes, four are soliloquies uttered by one figure, i.e. the Jailer's Daughter.

21 Sometimes not only the awareness of time and space changes, but also the awareness of one's "I": for instance when an actor starts using third person in reference to the figure he/she is impersonating, he/she is simply undermining illusion even further, for the change of grammatical *deixis* signals the distance separating the actor from the figure. However, also in this case the actor is performing a brief role of a fictional actor making an aside.

22 The theatrical freeze is a sign of the suspension of time flow within the fictional world (we are aware that the biological time of the actors flows with the same tempo as before). Since this is conspicuous and unconcealed evidence for the difference between the two time streams, the actors' and the figures', the conventionality of theatre is unveiled.

23 As States put it: "the actor who plays to the audience in the aside or the monologue is usually well within the play world, since the audience he addresses is only the idea of an audience" – op. cit., p. 171.

24 During an aside whom we see is (a) a real human being whose profession or hobby is acting, (b) the fictional figure he/she is enacting, being in this particular case an actor, and (c) the figure that the "fictional" actor enacts who is both the "real" actor and the figure he/she has been hitherto enacting. In other words, as an icon, the real actor ceases to be a sign of someone else, but becomes a sign of him-/herself during a stage employment, and this "new" actor (also fictional) is the one who utters an aside.

25 This is also the feature of stage speech in postdramatic theatre.

26 The other possibility is that the actor has ceased playing Puck and now appears as himself only, a real human being. In this case, if he also spoke out of himself, theatricality would be lost completely and we would be dealing with a sort of one-man show.

27 A direct address to the audience by chorus or narrator will be different in nature because figures of that kind not only are not heard but also are not seen by other figures, so their ability to see and address the spectators is not a violation of a boundary separating the two sides. By convention both chorus and narrator seem to belong to the same reality as the spectators and not the figures on stage. Again, however, their utterances are part of the script and reveal the fact that they are

234   *Notes*

not independent, but are in fact an inseparable part of a superior structure of a dramatic text, which is, of course, only one of the systems employed in theatre.

9   Theatre-Within-the-Theatre, or Time-Within-the-Time

1   Manfred Pfister, *The Theory and Analysis of Drama* (Cambridge: Cambridge University Press, 1988), p. 223.

2   In this chapter I have used parts of my essay "The Play-Within-The-Play: A Theoretical Perspective", *Enjoying the Spectacle: Word, Image, Gesture*, edited by Jerzy Sobieraj and Dariusz Pestka, Festschrift for Professor Marta Wiszniowska (Toruń: Wydawnictwo UMK, 2006), pp. 17–32.

3   A play that deals specifically with this particular convention is Tom Stoppard's *Rosencrantz and Guildernstern Are Dead*. For a thorough discussion of this play see Joanna Kokot, "'All the world's a stage': Theatre-within-Theatre Convention in Tom Stoppard's *Rosencrantz and Guildenstern Are Dead*", in: *Conventions and Texts*, edited by Andrzej Zgorzelski (Gdańsk: Wydawnictwo Uniwersytetu Gdańskiego, 2003), pp. 113–39.

4   In some plays, at least some of the fictional spectators (who sometimes are also the actors in the play within) are aware of the relationship between an element of the plot of the play staged and some events in their own reality (as is the case in *The Spanish Tragedy*, *Hamlet* or *The Roman Actor*).

5   In many films on theatre the stress is put on (a), i.e. on the indexical feature of acting by which the actor, while playing and impersonating someone else, shows his/her personal emotions (joy, satisfaction, pain, fear and the like) connected more with the fact of acting than with the fictional reality of the enacted figure. This may be exemplified by the last scenes in *Shakespeare in Love*, where Viola by chance finds herself on the stage playing Juliet. She is presented as delighted and also very much in love with Romeo and with Shakespeare who plays Romeo. So, the actress had a difficult task to show indexically Juliet's and her own emotions, which do not necessarily overlap. Juliet loves Romeo, whereas Viola loves the actor playing Romeo.

6   Please note that the rule of the juxtaposition of the two perspectives of perceiving reality (by the figures and the spectators) is still preserved here. The fictional spectators on the stage notice and perceive the actors on stage as real human beings who assume fictional roles. The real spectators see them as real human beings who assume the roles of fictional actors who, in turn, assume the roles of whoever they enact.

7   Manfred Pfister gives an example from Ludwig Tieck's play, *Der gestiefelte Kater* [*Puss in Boots*], in which the author allows the "Hanswurst" figure in the play within to address the real audience "as a dialogue partner and has the fictional audience, Fischer and Müller – who imagine that they are the ones being addressed – react to his comments with a complete lack of comprehension", which creates a comic discrepancy of awareness between the figures and the real audience. Cf. Pfister, op. cit., p. 226. Fischer and Müller are not aware of the theatrical situation they are involved in themselves, and do not notice the existence of the real audience. This also means that a sudden break in the illusion is created, for Hanswurst has the ability to see through the fictionality, and address the real audience directly, a signal that is only partially perceived by the two figures in the macro play.

8   I have used Colin Gibson's edition (Cambridge: CUP, 1978).

9   Even the music is external to the dumb show, as evidenced by Conjurer's own words: "Strike louder music from this charmed ground,/ To yield, as fits the act, a tragic sound" (2.2.36–7).

10 It has been noted that Webster was not the first dramatist to use this device. Robert Greene, for instance, employed a dumb show to present events taking place somewhere else in his *Friar Bacon and Friar Bungay* (1589).

11 In what follows I have partly used the material contained in my essay "Thomas Kyd's *The Spanish Tragedy*: A Play About a Play", *Studia nad literaturami europejskimi. Księga poświęcona pamięci Profesora dr hab. Henryka Zbierskiego* (Poznań: Motivex, 1999), pp. 115–25.

12 On classical analogues of this episode see Eugene D. Hill's "Senecan and Virgilian Perspectives in *The Spanish Tragedy*", *ELR*, 15 (1985), pp. 143–63.

13 I am using Philip Edwards' edition throughout this essay (The Revels Plays series, Manchester University Press, 1977). For a discussion of *The Spanish Tragedy's* theology see Philip Edwards' essay, "Thrusting Elysium into Hell: The Originality of *The Spanish Tragedy*", *The Elizabethan Theatre XI* (Ontario: P. D. Meany, 1990), pp. 117–32.

14 In Edwards' edition I have come across only one "aside" at 3.13.56, but even that does not need to be addressed to the stage audience directly.

15 As John Scott Colley has noticed in his article "*The Spanish Tragedy* and the Theatre of God's Judgement", *Papers on Language and Literature*, 10 (1974): "By constantly introducing shows that have hidden truths for their audiences, Kyd emphasises the ways in which theatrical illusions reflect more levels of truth than may be apparent. Throughout the play occur little dramas, pageants, and emblematic presentations with obscure meanings that must be clarified by their several authors. These 'little plays' demonstrate how the audience is to react to the larger play that includes Hieronimo as well as Revenge" (p. 244).

16 Playing in different languages has become a vogue in recent times. See Marvin Carlson, *Speaking in Tongues,* op. cit.

17 On the apocalyptic aspect of the play see Frank Ardolino's "'Now Shall I See the Fall of Babylon': *The Spanish Tragedy* as Protestant Apocalypse", *Shakespeare Yearbook,* 1 (Spring 1990), pp. 93–15.

Bibliography

And, Metin, *Drama at the Crossroads. Turkish Performing Arts Link Past and Present, East and West* (Istanbul: The Isis Press, 1991)

Arden, John, *Three Plays* (Harmondsworth: Penguin, 1964)

Auslander, Philip, *Liveness* (London and New York: Routledge, 1999)

Avigal, Soshana and Shlomith Rimmon-Kenan, "What do Brook's Bricks Mean?", *Poetics Today*, 3.3 (Spring 1981), pp. 11–34

Banes, Sally and André Lepecki (eds), *The Senses in Performance* (London and New York: Routledge, 2007)

Barnett, Dene, *The Art of Gesture: The Practices and Principles of 18ᵗʰ Century Acting* (Heidelberg: Carl Winter Universitätsverlag, 1987)

Baugh, Christopher, *Theatre, Performance and Technology. The Development of Scenography in the Twentieth Century* (Basingstoke and New York: Palgrave Macmillan, 2005)

Beckermann, Bernard, *Theatrical Presentation. Performer, Audience and Act* (New York and London: Routledge, 1990)

Carlson, Marvin, "The Haunted Stage: Recycling and Reception in the Theatre", *Theatre Survey*, 35.1 (Spring 1994), pp. 5–18

Carlson, Marvin, *Speaking in Tongues. Languages at Play in the Theatre* (Ann Arbor: The University of Michigan Press, 2006)

Carroll, Noël, "Philosophy and Drama. Performance, Interpretation, and Intentionality", in: *Staging Philosophy. Intersections of Theater, Performance, and Philosophy*, ed. David Krasner and David Z. Saltz (Ann Arbor: The University of Michigan Press, 2006), pp. 104–121

Clark, Herbert H. and Edward F. Schaefer, "Dealing with overhearers", in: Herbert H. Clark (ed.) *Arenas of Language* (Chicago: University of Chicago Press, 1992), pp. 248–74

Cook, Amy, "Interplay: The Method and Potential of a Cognitive Scientific Approach to Theatre", *Theatre Journal*, 59.4 (2007), pp. 579–94

Daileader, Celia R., *Eroticism on the Renaissance Stage. Transcendence, Desire, and the Limits of the Visible* (Cambridge: Cambridge University Press, 1998)

De Certeau, Michel, "Vocal Utopias: Glossolalias", *Representations*, 56 (Fall 1996), s. 29–47

De Marinis, Marco, *The Semiotics of Performance* (Bloomington and Indianapolis: Indiana University Press, 1993)

De Marinis, Marco, *From Script to Hypertext: Mise en Scéne and the Notation of Theatrical Production in the Twentieth Century*, *Gestos*, 23 (April 1997), pp. 9–37

De Toro, Fernando, *Theatre Semiotics. Text and Staging in Modern Theatre* (Toronto and Buffalo: University of Toronto Press, 1995)

Dillon, Janette, "Performance Time: Suggestion for a Methodology of Analysis", *Medieval English Theatre*, 22 (2000), pp. 33–51

Dixon, Steve, *Digital Performance. A History of New Media in Theater, Dance, Performance Art, and Installation* (Cambridge, Mass., London: MIT Press, 2007)

Dorval, Patricia, *Towards a Stylistics of the Modes of Ostension*, *Theatre Research International*, 18.3 (1993), pp. 206–14

Edwards, Philip, "Thrusting Elysium into Hell: The Originality of *The Spanish Tragedy*", *The Elizabethan Theatre XI* (Ontario: P.D. Meany, 1990), pp. 117–32

Evans, Vivyan, *The Structure of Time. Language, meaning and temporal cognition* (Amsterdam/Philadephia: John Benjamins Publishing Company, 2004)

Fischer-Lichte, Erika, "Introduction: Theatricality: A Key Concept in Theatre and Cultural Studies", *Theatre Research International*, 20.2 (1995), pp. 85–9

Fischer-Lichte, Erika, *The Semiotics of Theater*, translated by Jeremy Gaines and Doris L. Jones (Bloomington and Indianapolis: Indiana University Press, 1992)

Fischer-Lichte, Erika, *Estetyka performatywności* [*Ästhetik des Performativen*], translated by Mateusz Borowski and Małgorzata Sugiera (Kraków: Księgarnia Akademicka, 2008)

Fleck, Andrew, "'Ick verstaw you niet': Performing Foreign Tongues on the Early Modern English Stage", *Medieval and Renaissance Drama in England*, 20 (2007), pp. 204–21

Garner, Stanton B. Jr., *Bodied Spaces. Phenomenology and Performance in Contemporary Drama* (Ithaca and London: Cornell University Press, 1994)

Giesekam, Greg, *Staging the Screen. The Use of Film and Video in Theatre* (Basingstoke and New York: Palgrave Macmillan, 2007)

Goffman, Erving, *The Presentation of Self in Everyday Life* (London: Penguin, 1959)

Gordon, Robert, *The Purpose of Playing. Modern Acting Theories in Perspective* (Ann Arbour: The University of Michigan Press, 2006)

Grabes, Herbert, "Staging plays in the theatre of the mind", in: *Reading Plays. Interpretation and Reception*, ed. Hanna Scolnicov and Peter Holland (Cambridge: Cambridge University Press, 1991), pp. 94–109

Green, J. R., *Theatre in Ancient Greek Society* (London and New York: Routledge, 1994)

Hill, Eugene D., "Senecan and Virgilian Perspectives in *The Spanish Tragedy*", *ELR*, 15 (1985), pp. 143–63

Hirsh, James, *Shakespeare and the Theory of Soliloquies* (Madison: The University of Delaware Press, 2003)

Hortmann, Wilhelma, *Word into Image: notes on the scenography of recent German productions*, in: *Foreign Shakespeare. Contemporary Performance,* ed. Dennis Kennedy (Cambridge: Cambridge University Press, 1993), pp. 232–53

Ichikawa, Masiko, "What to do with a Corpse?: Physical Reality and the Fictional World in the Shakespearean Theatre", *Theatre Research International*, 29.3 (2004), pp. 201–215

Jacob, Pierre and Marc Jeannerod, *Ways of Seeing: The Scope and Limits of Visual Cognition* (Oxford: Oxford University Press, 2003)

Jakobson, Roman, "O językoznawczych aspektach przekładu", in: *W poszukiwaniu istoty języka*, ed. Maria Renata Mayenowa, vol. 1 (Warsaw: PIW, 1989)

Johnson, Mark, *The Meaning of the Body. Aesthetics of Human Understanding* (Chicago & London: The University of Chicago Press, 2007)

Jones, Amelia and Andrew Stephenson (eds), *Performing the Body/Performing the Text* (London and New York: Routledge, 1999)

Kiernan, Pauline, *Shakespeare's Theory of Drama* (Cambridge: Cambridge University Press, 1996)

Kirby, Michael, "On Acting and Non-Acting", in: *Acting (Re)Considered. A Theoretical and Practical Guide*, ed. Phillip Zarrilli (London and New York: Routledge, 2002), pp. 40–52

Kokot, Joanna, "'All the world's a stage': Theatre-within-Theatre Convention in Tom Stoppard's *Rosencrantz and Guildenstern Are Dead*', in: *Conventions and Texts*, ed. Andrzej Zgorzelski (Gdańsk: Wydawnictwo UG, 2003), pp. 113–39

Konijn, Elly A., *Acting Emotions. Shaping Emotions on Stage* (Amsterdam: Amsterdam University Press, 2000)

Konijn, Elly, "Actor's Emotions Reconsidered: A psychological task-based perspective", in: *Acting (Re)Considered. A theoretical and practical guide*, ed. Phillip B. Zarrilli (London and New York: Routledge, 2002), pp. 62–81

Lakoff, George and Mark Johnson, *Philosophy in the Flesh. The Embodied Mind and Its Challenge to Western Thought* (New York: Basic Books, 1999)

Limon, Jerzy, *Dangerous Matter. English Drama and Politics in 1623/24* (Cambridge: Cambridge University Press, 1986)

Limon, Jerzy, *The Masque of Stuart Culture* (Newark, Del.: The University of Delaware Press, 1990)

Limon, Jerzy, "From Liturgy to the Globe: the Changing Concept of Space", *Shakespeare Survey*, 52 (1999), pp. 46–53

Limon, Jerzy, "Thomas Kyd's *The Spanish Tragedy*: A Play About a Play", *Studia nad literaturami europejskimi. Księga poświęcona pamięci Profesora dr hab. Henryka Zbierskiego* (Poznań: Motivex, 1999), pp. 115–25

Limon, Jerzy, "Boska perspektywa: czas i przestrzeń w dramacie liturgicznym", in: *Paradoksy humanistyki. Księga pamiątkowa ku czci Profesora Andrzeja Zgorzelskiego*, ed. Olga Kubińska and David Malcolm (Gdańsk: Wydawnictwo Uniwersytetu Gdańskiego, 2001)

Limon, Jerzy, "Shakespeare the Semiotician, or, Shakespeare Writes about His Own Art", in: *Playing Games with Shakespeare*, ed. Olga Kubińska and Ewa Nawrocka (Gdańsk: Theatrum Gedanense Foundation, 2005), pp. 107–19

Limon, Jerzy, "Shakespeare's Soliloquies and Asides: A Theoretical Perspective", *Kwartalnik Neofilologiczny*, LII.4 (2005), pp. 301–23

Limon, Jerzy, "Kto trzyma kaleidoskop, czyli o zalążkach terroru", *Dialog*, 10 (2006), pp. 100–11

Limon, Jerzy, "Kneblowanie teatru", *Notatnik Teatralny*, 45–46/(2007), pp. 171–80

Limon, Jerzy, *Piąty wymiar teatru* (Gdańsk: słowo obraz/terytoria, 2006)

Limon, Jerzy, "The Play-Within-The-Play: A Theoretical Perspective", in: *Enjoying the Spectacle: Word, Image, Gesture*, ed. Jerzy Sobieraj and Dariusz Pestka, Festschrift for Professor Marta Wiszniowska (Toruń: Wydawnictwo UMK, 2006)

Limon, Jerzy, "The Fifth Wall: Words of Silence in Shakespeare's Soliloquies and Asides", *Shakespeare Jahrbuch*, 144 (2008), pp. 47–65

Limon, Jerzy, "Waltzing in Arcadia", *New Theatre Quarterly* (2008), pp. 222–8

Limon, Jerzy, "Archaeology of Memory", *Cahiers Elisabethains*, 73 (Spring 2008), pp. 39–47

Limon, Jerzy, "Rozbieranie (się) teatru", in: *Inna Scena. Ciało, Płeć, Pożądanie. Tożsamość seksualna i tożsamość płci w polskim dramacie i teatrze*, ed. Agata Adamiecka-Sitek and Dorota Buchwald (Warsaw: Instytut Teatralny, 2008), pp. 41–58

Limon, Jerzy, "Space is Out of Joint: Experiencing Non-Euclidean Space in Theatre", *Theatre Research International*, 33.2 (2008), pp. 127–44

Limon, Jerzy, "A Candle of Darkness: Multiple Deixis in Roberto Ciulli's *King Lear*", *Journal of Dramatic Theory and Criticism* (Spring 2008), pp. 83–102

Limon, Jerzy, "Theatre's Fifth Dimension: Time and Fictionality", *Poetica*, 41, H. 1–2 (2009), pp. 33–54

Limon, Jerzy and Agnieszka Żukowska (eds), *Theatrical Blends* (Gdańsk: Theatrum Gedanense Foundation, 2010)

Lopez, Jeremy, "Imagining the Actor's Body on the Early Modern Stage", *Medieval and Renaissance Drama in England*, 20 (2007), pp. 187–203

Loxley, James, *Performativity* (London and New York: Routledge, 2007)

McAuley, Gay, *Space in Performance: Making Meaning in the Theatre* (Ann Arbor: The University of Michigan Press, 1999)

McConachie, Bruce, "Falsifiable Theories for Theatre and Performance Studies", *Theatre Journal*, 59.4 (2007), pp. 553–77

Mancing, Howard, "See the play, read the book", in: *Performance and Cognition. Theatre studies and the cognitive turn*, ed. Bruce McConachie and F. Elizabeth Hart (London and New York: Routledge, 2007) pp. 189–206

Martin, Jacqeline, *Voice in Modern Theatre* (London and New York: Routledge, 1991)

Meagher, John C., *Shakespeare's Shakespeare. How the Plays Were Made* (New York: Continuum, 1997)

Millar, Oliver, *Rubens: The Whitehall Ceiling* (London: Oxford University Press, 1958)

Nanay, Bence, "Perception, Action, and Identification in the Theater", in: *Staging Philosophy. Intersections of Theater, Performance, and Philosophy*, ed. David Krasner and David Z. Saltz (Ann Arbor: The University of Michigan Press, 2006), pp. 244–54

Oddey, Alison and Christine White (eds), *The Potentials of Spaces. The History and Practice of Scenography and Performance* (Bristol: Intellect Books, 2006)

Paivio, Allan, *Mental Representations: A Dual Coding Approach* (New York: Oxford University Press, 1986)

Pavis, Patrice, *Dictionary of the Theatre: Terms, Concepts and Analysis*, translated by Christine Schantz (Toronto: University of Toronto Press, 1998)

Pavis, Patrice, *Analyzing Performance. Theatre, Dance and Film* (Ann Arbor: The University of Michigan Press, 2003)

Pfister, Manfred, *The Theory and Analysis of Drama* (Cambridge: Cambridge University Press, 1988)

Pinto, J. C. M., *The Reading of Time. A Semantico-Semiotic Approach* (Berlin and New York: Mouton de Gruyer, 1989)

Reid, Francis, *The Stage Lighting Handbook* (London: A&C Black, 2001)

Rozik, Eli, "The Vocabulary of Theatrical Language", *Assaph*, Section C, No. 2 (1985), pp. 15–26

Rozik, Eli, "The Syntax of Theatrical Communication", *Assaph*, Section C, No. 3 (1986), pp. 43–57

Rozik, Eli, "The Functions of Language in the Theatre", *Theatre Research International*, 18.2 (1993), pp. 104–14

Rozik, Eli, "Categorization of Speech Acts in Play and Performance Analysis", *Journal of Dramatic Theory and Criticism*, VIII.1 (Fall 1993), pp. 117–32

Rozik, Eli, "The Corporeality of the Actor's Body: The Boundaries of Theatre and the Limitations of Semiotic Methodology", *Theatre Research International*, 24.2 (1999), pp. 198–211

Rozik, Eli, "Is the Notion of 'Theatricality' Void?", *Gestos*, 15.30 (November 2000), pp. 11–30

Rozik, Eli, "A Silent Theatrical Production Representing a Speaking Fictional World: Analysis of 'A Fleeting Shadow'", *Theatre Research International*, 27.1 (2002), pp. 78–96

Rozik, Eli, "Acting: The Quintessence of Theatricality", *SubStance*, 31.2 & 3 (2002), pp. 110–124

Rozik, Eli, *The Roots of the Theatre: Rethinking Ritual and Other Theories of Origin* (Iowa City: University of Iowa Press, 2002)

Rozik, Eli, *Generating Meaning in Theatre* (Brighton and Portland: Sussex Academic Press, 2008)

Saltz, David Z., "What Theatrical Performance Is (Not): The Interpretation Fallacy", *Journal of Aesthetics and Art Criticism*, 59.3 (Summer 2001), pp. 299–306

Schechner, Richard, *Performance Theory* (New York and London: Routledge, revised edition 1988)

Schlueter, Kurt, "Introduction to TGV," *The New Cambridge Shakespeare* (Cambridge: Cambridge University Press, 1990)

Shakespeare, William, *All's Well That Ends Well*, The New Cambridge Shakespeare, ed. Russell Frase (Cambridge: Cambridge University Press, 1985)

Shepherd, Simon, *Theatre, Body and Pleasure* (London and New York: Routledge, 2006)

Simon, E. (ed.), *The Theatre of Medieval Europe. New research in Early Drama* (Cambridge: Cambridge University Press, 1991)

Sofer, Andrew, *The Stage Life of Props* (Ann Arbor: The University of Michigan Press, 2003)

States, Burt O., *Great Reckonings in Little Rooms. On the Phenomenology of Theater* (Berkeley, Calif.: University of California Press, 1987)

States, Burt O., "The Actor's Presence. Three Phenomenal Modes", in: *Acting (Re)Considered. A Theoretical and Practical Guide*, ed. Phillip B. Zarrilli (London and New York: Routledge, 2002), pp. 23–39

Strong, Roy C., *Britannia Triumphans. Inigo Jones, Rubens and Whitehall Palace* (London: Thames and Hudson, 1980)

Yi-Fu Tuan, *Przestrzeń i miejsce* (Warsaw: PIW, 1987)

Turner, Cathy and Synne K. Behrndt, *Dramaturgy and Performance* (Basingstoke and New York: Palgrave Macmillan, 2008).

Veltruský, Jiří, "Contribution to the Semiotics of Acting", in: *Sound, Sign and Meaning. Quinquagenary of the Prague Linguistic Circle*, ed. Ladislav Matejka (Ann Arbor: The University of Michigan Press, 1978), pp. 553–606

Veltruský, Jiří, "Acting and Behaviour: A Study of the *Signans*", in: *Semiotics of Drama and Theatre. New Perspectives in the Theory of Drama and Theatre*, ed. Herta Schmid and Aloysius Van Kesteren (Amsterdam/Philadelphia: John Benjamins Publishing Company, 1984), pp. 393–441

Weber, Samuel, *Theatricality as Medium* (New York: Fordham University Press, 2004)

West, William N., *The Idea of a Theater: Humanist Ideology and the Imaginary Stage in Early Modern Period*, Renaissance Drama, New Series XXVIII, *The Space of the Stage* (Evanston: Northwestern University Press, 1999), pp. 245–87

Whitmore, Jon, *Directing Postmodern Theater. Shaping Signification in Performance* (Ann Arbor: The University of Michigan Press, 1994)

Zich, Otakar, *Estetika dramatického umění* (Praha, 1931)

Index